SCIENCE AND SUPERSTITION

SCIENCE AND SUPERSTITION

BY

M. N. ROY

GYAN

Published by

Gyan Publishing House
5, Ansari Road
Daryaganj, New Delhi-110002
Phone: 011-47034999, 9811692060
E-mail: books@gyanbooks.com

Distribution Network
gyanbooks.com
India, USA, Canada, UK, Australia

ISBN: 978-93-6280-055-8 (HB)
First Published, 1940

2nd Impression 2024

Printed at: Gyan Press, Delhi.

SCIENCE AND SUPERSTITION
Author: M. N. ROY

SCIENCE AND SUPERSTITION

M. N. Roy

CONTENTS

PAGE

PREFACE ... 1

CHAPTER

I. THE TRANSMIGRATION OF SOUL ... 5

II. PSYCHOLOGY OF THE SEER ... 55

III. CRIME AND KARMA 103

IV. WHY MEN ARE HANGED ... 114

PREFACE

ALTHOUGH this book is a collection of random notes, made in jail, it can be regarded as an introduction to the study of two important branches of social science, namely, criticism of the religious mode of thought and criminology. The transmigration of soul and the law of *karma* are the fundamental articles of faith with the vast bulk of the Indian people. The entire religious mode of thought, which still dominates the spiritual and intellectual atmosphere of our country, is rested on those twin-pillars. Modern education and penetration of scientific knowledge are challenging the religious mode of thought. Yet, prejudice dies hard. The efforts made even by people with modern scientific education to rationalise the religious mode of thought is only a matter of prejudice. A criticism of religious thought, subjection of traditional beliefs and the time-honoured dogmas of religion to a searching analysis, is a condition for the belated Renaissance of India. The spirit of enquiry should overwhelm the respect for tradition. This book is expected to quicken that spirit.

Superstition is rooted in the ignorance of the primitive man. In course of time man outgrows the blissful state of ignorance. Nevertheless, he is haunted by superstitions haloed by tradition, and often raised to the dignity of the expression of revealed wisdom. Eventually, scientific knowledge gives him the power to break the spiritual bondage. The history of the development of science coincides with the history of a bitter struggle against superstition. In our country, the struggle is still to begin. Whatever little of modern scientific knowledge is now there, is very largely superficial, and is often utilised with the purpose of reinforcing superstitions. That is an abuse of science.

This book is bound to provoke an outburst of criticism. But that will not be serious criticism; it will be an arrogant condemnation of the scientific spirit and scientific knowledge. At the same time, the purpose of initiating an organised struggle against superstition will be served. The clay-feet of a number of time-honoured gods are exposed in this book. Fatalism and blind faith have killed in the bulk of the Indian people the incentive for knowledge and progress. The root of this evil can be traced to the doctrine of the transmigration of soul.

Therefore, the exposure of the fallacy of this doctrine is a historical necessity. It is necessary not only for the material progress, but also for the spiritual liberation of the Indian people.

A critique of the cult of "religious experience" is equally necessary. That requires not only some knowledge of modern psychology, but a good deal of moral courage. Because, in the prevailing intellectual atmosphere of our country, it amounts almost to heresy. How superstition treats the heretic, is a dreadful tale. Nevertheless, the heretics are harbingers of real spiritual progress. In this book, the psycho-pathological foundation of the cult of "religious experience" has been exposed.

The last two chapters of the book are parts of my prison memoirs. So very strong is still the grip of superstition on the mind of the Indian masses that it affects crime and punishment also. The so-called criminals, owing to the cruel treatment meted out to them, usually are rebels against society. But that is not the case in India. Superstition persuades the victims of social tyranny to accept the most cruel punishment with resignation, firmly believing that the entire responsibility belongs to themselves. This is illustrated by the record of a number of murder cases. All of them reveal the sordidness of the social background of the crimes. It is hoped that the last two chapters of the book will serve as the incentive for the study of criminology as a part of social science.

On the whole, the book is written with the purpose of provoking thought. It is too small a work to do more than indicate the scientific approach to the difficult problem of overcoming the age-long tyranny of superstition.

Dehradun, July 1940. M. N. Roy

CHAPTER I

THE TRANSMIGRATION OF SOUL

THERE is a new excitement among the inmates of this little world of ours. Not exactly an excitement. It is rather a commotion—a futile flutter. Excitement is an emotion caused by events which directly affect ourselves intimately. Nothing like that has happened. Only a story from beyond the walls has filtered in, and everybody is repeating it with some additional embellishing touches. It is surprising how stories do reach us in this segregated world with all its paraphernalia of rigid watch and ward. However, whenever they break through the blockade, they spread like wild fire, to the edification of solitary souls hungry for information. Particularly, when the stories are of the nature that stimulates idle fancy or feeds general credulity. Some of the warders would pick one up in the neighbouring bazar, and pass it on in a distorted or magnified form. Or some prisoners employed in the office might get snatches of some conversation among the clerks.

The particular story causing the present commotion seems to have emanated from this source, which is usually regarded by the prisoners as beyond all possible doubt. It is said to be causing a little storm even in the greater world beyond the walls. It is about a girl remembering the events of her past life so vividly as to give yet another knock-out blow to those who, corrupted by the influence of the materialist West, question the truth of the transmigration of soul.

The story comes from holy Muttra. Can you possibly disbelieve anything that is reported to have happened in that place where countless miracles were performed by Lord Krishna when he was a naughty boy or an amorous adolescent? Here is the story as I could piece it together. It is usually related many times a day, accompanied with so many pious exclamations that, to get at what is supposed to be its sub-stratum of fact, is extremely difficult. It is marvellous how tenacity of belief can become the measure of accuracy; and the blinder the faith, the more tenacious it is.

The centre of the story, Shanti Devi, was born, presumably for the millionth time or thereabouts, nine years ago, the daughter of a Brahmin resident of a place in the neighbourhood of Delhi. It is reported that, for some time, she has been urging her parents to take her to Muttra where she claims to have lived her previous life. In the beginning, the parents, so they say, did not pay any heed to her assertion and request. But presently her claim became known to others outside the family. A number of local gentlemen formed themselves into a "Committee of Investigation," and took the girl to Muttra. There she is reported to have performed the following miraculous acts. On the station platform, she recognised a local Brahmin resident as her husband of the former life; on the way, from a crowd of spectators, she picked out an old gentleman as her previous father-in-law; she directed the Investigating Committee to the house she had lived in her last life; she showed intimate acquaintance with the lay-out of the house; she indicated the ghat on the Jumna where she used to bathe; she, of course, manifested great tenderness to the young son of her former husband; the latter reported that his wife had died nine years ago at the birth of that boy. Now, in the face of all these facts, who can doubt the truth of the transmigration of soul?

The Committee, composed of gentlemen firmly convinced of the truth, naturally, had no difficulty in coming to the conclusion that Shanti Devi's story had been completely borne out by facts. The conclusion was reported publicly in the press and from the platform. Shanti Devi herself appeared in public meetings, and related her story which was then corroborated by one or the other member of the Investigating Committee. Enthusiastic defenders of Hindu culture rushed to the press jeering at the pretensions of modern science and ridiculing the hocus-pocus of new-fangled psychology. Science was dared to take up the challenge thrown down by the nine-years old heroine who has never heard of such strange names as physics and psychology; her acquaintance with the three R's is most probably still to be made, if she would ever be contaminated by these symbols of wicked worldliness. The aggressive apostles of cultural nationalism themselves are rather subjects of psychiatrical and psycho-analytical study than competent critics of modern science.

Having heard the story repeated time and again, I naturally turned it over and over in my mind. What interested me, was not so much the question of transmigration. I am indifferent to the question. Once upon a time, man had reason to invent a soul; they must believe in their own invention. There are those, who, having known better, are scornful of antiquated toys. Their disbelief is vehemently condemned or loftily deplored by others. I do not share the belief in truths *invented* by man when he was incapable of *discovering* objective truths. Truths discovered are a different matter. Being physical facts, they belong to the world of reality. I define truth as a physical fact. That requires a word of explanation. The term "physical" includes biological as well as mental. This is a sound statement incontestable scientifically. Metaphysicians may resent it; but resentment or dogmatic assertion is no logic. Man's gods are made after their own image. I have no more respect for gods than for their makers. Have I not seen the clay-feet of the makers of gods? So, the story of Shanti Devi did not stimulate me to meditate over the truth of transmigration. I simply did not believe in it, having no reason to fictionalise the fact of my existence. Men are perverted enough to be ashamed of being what they are,—animals. Being thinking animals, they should have more sense; but most of them don't. So much the worse for them. To possess the faculty of thought, but not to use it, is a misfortune. Therefore, they pretend to be gods—try to hide the facts of their being with a fiction fabricated by perverted imagination. They prefer falsehood to truth.

The subject of my reflection was the credulity on the part of people expected to be more discerning and discriminating. In Shanti Devi's meetings, there sat men and women who could not possibly be blind to the obvious perfunctoriness of the enquiry into the phenomenon; who could easily raise a whole host of questions that are to be answered before any conclusion could be accepted even hypothetically.

The fundamental question is about the reliability of the evidence. If one simply believes in the transmigration of soul, there is nothing to be said about it. You cannot argue about an article of faith. One comes to believe in transmigration through the simplest process of thought, if it can be placed under that category of mental act which involves reason: Begin with the assumption that there is a soul which transcends the biological being of man. In the strict scientific language, the term physical would be adequate, because the physical being embraces everything that really exists; however, I use the term biological to make it explicit that the soul of religion is not the sum total of the intellectual and emotional activities of man. Having begun with an assumption which, by its very nature, can never be experimentally established, you further assume that soul is immortal. This additional assumption is necessary to differentiate the imaginary essence of man from the biological reality of his being. As the immortal essence of man, soul naturally survives physical death. And being a disembodied spirit, its tangible immortality can be realised only in the acquisition of a new corporal abode every time it is rendered homeless by the relentless operation of biological laws which it cannot control. Thus, the dogma of transmigration is deduced from the assumption of an immortal essence of man, which assumption, in the last analysis, has no other basis than the primitive animistic notion of a world spirit.

The idea of an immortal soul is not the acquisition of man when he reaches a high spiritual level. It is a very very primitive idea, its origin being not spiritual elevation, but ignorance. The phenomenon can be observed even now among the primitive races. The aboriginal inhabitants of the Malay Peninsula believe that souls are red, no bigger than grains of maize; for other Malay races, they are vapoury, shadowy, filmy essences, about as big as one's thumb[1]; in other parts of the Pacific Islands, soul is conceived of not as a tiny being confined to a single part of the body, but as a sort of fluid diffused through every part; the backward masses of Japan considered the soul as a small, round, black thing; the Australian Bushman also believes the soul to be a small thing dwelling in the breast.[2]

Primitive people explain natural occurrences as caused by the action of spirits which are believed to appear and operate as (1) ghosts, that is, spirits which have formerly been incarnate; (2) dream-spirits which have temporarily left bodies in sleep or trance; (3) invisible, conscious beings which have never been incarnate. The conception of the object of this belief, common to the primitive man, wherever or whenever he may live, varies from place to place and time to time. But

---

[1] The *shukshma shareer* of the Hindu scriptures is believed also to be of the same size.

[2] Carveth Read, "Man and His Superstitions."

eventually, it divides itself in two distinct schools, so to say, which have been named by anthropologists as Hyperphysical Animism and Psychological Animism. One regards an object as being moved by the spirit inherent in itself; the other attributes all movements to an agent which possesses the object for the time being, but is separable from it. Hyperphysical Animism conceives consciousness as a distinct entity capable of quitting the body, surviving its death and existing independently as disembodied spirit. Psychological Animism, on the contrary, ascribes anthropomorphic consciousness to all objects, particularly to the animate. This type of Animism can be detected in children even among civilised peoples. They are seen to hit back at such inanimate things as a table or a chair upon being hurt by bumping against it. Evidently, the doctrine of soul and its transmigration evolved out of the Hyperphysical Animism which has been traced to the primitive man's desire to explain dreams, wherein the dead appears as in flesh.[1] The explanation is that on death the spirit leaves the body. Later on, there begins speculation about what the spirit is made of. The "soul-stuff" is conceived as material, though subtle and normally invisible. It is believed to be permeating the whole body.[2]

---

[1] E. B. Tylor, "Primitive Culture."

[2] J. G. Frazer, "Belief in Immortality."

"This conception of soul-stuff may have been an important contribution to metaphysics. The doctrine of material substance is reached by abstracting all the qualities of things; but then, there would be nothing left, were it not for this venerable idea of something invisible and intangible in things in which qualities may inhere, or which may serve as a support to them .... Along another line of speculation, this soul-stuff may become the Soul of the World when by philosophers spirits are no longer conceived to have bodies, but to be the very opposite of bodies, a spiritual substance must be invented to support their qualities, in order to put them upon an equal footing with reality with corporeal things .... but such speculations are confined to philosophers and theologians: some of whom maintain (as if reverting to the original savage idea) that spirit is the true substance of material things, at least that material things depend upon a spirit, or spirits, for their existence."[1]

For ages, the belief in ghosts remained mixed up in popular mind with the idea of the "soul-stuff". Eventually, metaphysical subtleties about the difference between mind and matter, spirit and body, conceived of the notion of a pure incorporeal, immortal spirit. The doctrine of transmigration was a logical outcome of that notion. However, even great modern metaphysi-

[1] Carveth Read, "Man and His Superstitions."

cians bound, for their own prestige as philosophers, to have some regard for rationalism and scientific knowledge, have to admit that the venerable doctrine of soul originates in the ignorance of primitive man. The famous German metaphysician Wundt came to the conclusion that the spirit of the living body is the starting point of Animism.[1]

The doctrine of soul being thus a spiritual relic of savagery, it may still hold its sway over the mind of the ignorant. But it is an entirely different proposition to claim scientific or even empirical support for the doctrine of transmigration. Whatever may be the origin of the soul, its process of transmigration takes place here and now. If it is an objective reality, there must be some way of observing it. Deny the reality of scientific knowledge, dispute the validity of the scientific mode of thought, and science cannot do anything but to leave you with your blind faith. But try to rationalise the dogmas of religion by claiming empirical basis for venerable superstitions, and you tread on a dangerous ground. If you challenge science, then, the issues thus joined must be fought out scientifically; you must observe the rules of the game. Scientific mode of thought differentiates itself from the religious mode of thought by refusing to accept any unverifiable hypothesis as the premise for

[1] "Myths and Religion."

deductions claiming to be objectively true. Scientific and rationalist thought rejects religion not because its dogmas cannot stand the test of science and rationalism, but because of the fallacious nature of the religious mode of thought itself, because of its own internal contradictions. If the religious dare to fight science and rationalism on the latter's ground, then, the combat must be conducted according to the scientific and rationalist methods. If they undertake to adduce empirical evidence in support of their superstitions, so that these could claim superiority to scientific knowledge, then, certain elementary laws of evidence shall have to be observed. The method of collecting evidence must be such as guarantees reliability.

There was a statement by a young girl. It is claimed that the statement has been verified. The whole case rests on the assumption that the enquiry was conducted impartially. But is impartiality possible in such an enquiry? Preconceived notions rule out criticism; evidence given by the superstitious and recorded by the uncritical can never be relied upon. The enquiry into the case was not free from those defects. It was conducted by persons who regarded the case as yet another proof for their belief in the transmigration of soul. The evidence was given by people still less competent to participate in a scientific investigation. All concerned were ardent believers, untouched by the sceptical and critical spirit of enquiry which alone can lead to the discovery of objective truths.

It is not logically permissible to talk of proving the doctrine of transmigration empirically without showing that the assumption of immortal soul has even a hypothetical validity. The doctrine results from the belief in soul and its immortality. The ground of this belief is to be critically examined. Science challenges this belief, and has exposed its groundlessness. As far as it is concerned, the existence of soul (not in the broad sense, but in the definite religious sense) has to be proved before the question of transmigration can receive any serious consideration. Who bothers about the imaginary peregrination of a non-entity? If the existence of soul were not assumed, then, the enquiry into the phenomena supposed to indicate some existence after death would be of an entirely different character. The result of any enquiry is greatly determined by its point of departure.

The story of Shanti Devi, or any other similar tale[1], granted that its veracity has been empirically verified, does not necessarily prove the existence of an immortal soul. If it proves anything, it proves that memory survives death.

[1] There are any number of them told and generally believed. Since this particular story became the matter of public discussion, there have appeared in the press reports of other pretenders.

This is an extremely fallacious assertion, and particularly ruinous for the doctrine of soul. Memory is a biological function. It is stored up in the brain which is destroyed upon the death of the body. Loss of memory due to cerebral disorder is a well-known fact showing that memory depends on the normal functioning of the physiological apparatus called brain. And everything physiological is physical. The brain is a lump of organic matter which can be analysed into its chemical components. None has ever postulated the existence of disembodied brains. In the light of physiological knowledge, that would be absurd.

If memory survives death as an attribute of soul, what happens, then, to the idea of soul? It is materialised! To be possessed of a physiological property, soul must be a physical entity. But it is supposed to be a disembodied spirit, and the possibility of transmigration is deduced from that supposition. The soul has nothing in common with the body; it resides in a body, unaffected by its physical functions, as an unattached spectator; therefore, it does not die with the body. It discards a body like a set of old clothes, and goes on to enter a new, unless it has qualified for the state in which it can dispense with all clothes. Now we see that the old clothes stick to it; more correctly speaking, it does not leave them altogether. Because, if it did, there would be no memory. In any case, the smell of the old clothes clings to the soul, evidently spoiling its purity. One cannot really be unattached to something the memory of which is so very deeply impressed.

"Scientific proof" of the transmigration of soul thus destroys the very doctrine of soul. In order to be what it is believed to be, namely, disembodied spirit, soul should not have any memory. It is proposed to prove the existence of soul on the strength of its possessing a property which it should not possess, if it were what it is believed to be. In other words, trying to prove transmigration "scientifically," you only succeed in disproving the existence of a disembodied spirit. This is a curious procedure—like cutting the branch on which you sit.

The originators of the doctrine of soul postulated the *sukshma shareer* as the basis for the belief in transmigration. As Radhakrishnan says, "our ancients were courageous." They realised what transmigration involved, and moulded the concept of soul accordingly. The soul as conceived by them is not a disembodied spirit. It is of the size of a thumb. All these quaint and curious ideas are set forth at great length in the Upanishads. The ancients could be bold, because science was not yet to challenge their speculative assertions. Nor were these speculations altogether vain. They had a purpose—a very mundane one. The ancients were laying down the laws of the ruling class. A transcendental sanction was necessary.

So it was simply invented. The superstition of the savage was utilised for the purpose. The belief in transmigration had to be fostered for securing submission to the law of *karma.* The object of this is to defend the established social order—to keep everybody in his allotted place.

But to rationalise irrationalism, is a hopeless undertaking. An article of faith simply cannot be scientifically proved. Faith is above proof. What is believed to transmigrate, to survive death, is not a disembodied spirit, but the *sukshma shareer.* So, the doctrine of transmigration asserts that a small *replica of the body* survives death. If that is assumed, then, memory of past life is hypothetically possible. But here you are definitely on the enemy's ground, trapped with no hope of escape. You are making an assumption which is definitely of scientific nature, which can be submitted to an empirical test. If what survives death is a physical entity, there must be some way of finding out what happens to it before the next birth.

The assertion is not simply an affirmation of the indestructibility of matter or of the law of conservation. The death of the body, of course, is not the destruction of the matter composing it, but only chemical dissolution of this latter. Death is destruction of a morphological organisation of matter. Matter is immortal. But the dogma of *sukshma shareer* asserts that the *organisation of matter* survives—in miniature. There would be no sense in saying that matter survives death; because death occurs to an organisation of matter. Matter itself is not involved in the process. The idea of survival implies continuation after death of that which is supposed to be affected by the process. It implies survival of the organisation of matter. The body is destroyed in gross form, but an attenuated form of it, an exact replica of the organisation apparently destroyed, remains intact.

This is a proposition which can be put to scientific test. The process of a human organism dying can be observed in the minutest detail. Indeed, the nature of this process is already known fairly accurately. It is a matter of clinical observation. There is not the slightest evidence to show that at any point in the process a replica of the body leaves the dying organism. To carry over memory and the accumulated store of experience, that replica must have a physico-chemical structure, however fine; and such a structure cannot possibly elude scientific detection.

No, the ingenious doctrine of *sukshma shareer* won't do in this age when physiology has penetrated the mystery of death. It commanded credence in an atmosphere of ignorance about the structure of body and the nature of life. But without the assumption of the *sukshma shareer*, it is not possible to maintain that transmigration can be proved. From the point of view of scientific method, all stories like that of Shanti Devi, no matter how very "authoritatively" told and "scientifically" corroborated, are *prima facie* untenable. A disembodied spirit cannot have memory; and a physical organisation such as can carry over memory cannot survive death. On the one hand, you cannot prove your case on your own ground; on the other hand, when your proposition is apparently such as can be scientifically tested, it palpably rests on a false premise and therefore does not deserve further consideration.

The whole procedure is logically fallacious. It rests on the assumption of that which it seeks to prove. If survival after death is proved by "unchallengeable facts", then, immortality of soul and the existence of soul itself are proved. That is the argument. But the possibility of survival pre-supposes the existence of something which does not die with the body. If this pre-supposition were not there, all such stories would be received with extreme scepticism. Their veracity is *prima facie* doubtful from the point of view of scientific knowledge. In the absence of the presupposition, such stories could not be regarded *a priori* as corroborating an established doctrine, but as phenomena to be scientifically explained. An enquiry taken up with such an attitude of scepticism and scientific objectivity would adopt entirely different methods.

Recent psychological research shows that collecting evidence is not a simple affair at all. The reliability of evidence depends upon a large variety of factors which are independent of the moral integrity of the witness. Pre-occupation is the most decisive. Thanks to it, people imagine seeing things which are not seen by others not so pre-occupied. On the other hand, emotional or physical agitation makes people fail to notice the most obvious. The actual nature of the simplest and most obvious events is usually not fully realised by the great majority of those who happen to witness them. Only trained observers with keen intelligence are able to report events approximately correctly.

The pre-condition for a really scientific investigation is to ascertain that what are called facts are really facts. Without this preliminary caution, castles are built in the air only too easily, and are therefore bound to crash under the slightest impact of reason. Once fictions are taken for facts, the rest follows swimmingly. Many people who believe to have seen ghosts report an actual experience. But that does not prove that they have experienced any objective fact. Yet, such "facts"—that reliable persons have actually seen ghosts—persuade many intelligent people to believe in ghosts. Things imagined are seen by imaginative eyes. They are "facts," as long as the imagination lasts. Therefore, figments of imagination, though facts *as such*, cannot be taken for the evidence for any objective reality.

Let alone experimental psychology; anybody who has read cleverly written detective stories knows that observation is not an easy business. To see is not to observe. A few instances of experimental recording of evidence may benefit those who dogmatically assert that the "verification" of such stories as Shanti Devi's is a challenge to modern psychology. I am boldly or hopefully assuming that they can be benefitted, that they are not irreparably lost in prejudice.

W. M. Marston discovered the method of detecting through a mechanical indicator of blood-pressure, whether an arrested person on examination is telling the truth The following experiment was made by that famous American psychologist. The experimenter was in a room with eighteen other educated persons—mostly lawyers—who had no previous knowledge that there was going to be any experiment. They were all behaving spontaneously as in any chance gathering—all casually chatting. After a while, as arranged beforehand by the experimenter, a young man, dressed so as to attract attention, rushed into the room, and handed to the experimenter an yellow envelope. While the experimenter was occupied in reading the message, the bearer, as previously instructed, drew out a large knife in a way that everyone in the room could see him do so. Dr. Marston writes: "Not one of those eighteen witnesses noticed the knife! Their attention was on the supposed telegram. When asked in direct examination about the knife, they all denied seeing it. During cross-examination, they became still more vehement in their denials. They suspected that the cross-examiner was trying to trick them into making false statements. Yet the knife had been held in full view for approximately three minutes!"

Then there are the so-called Aussage-tests, which show how really difficult fact-finding is. Students in a class room are asked to write down everything they observe. The professor arranges for a variety of events to take place which the students are to report. The mistakes are amazing, simply incredible. Had not the same test been repeated over and over again in different colleges, with different groups of students, the story would sound fantastic. On the average, less than thirty per cent of the students report the events correctly. Such amazing mistakes, for example, are made: One of the events enacted was several people exchanging heated words. A number of students not only failed to report such an outstanding occurrence, but when asked about it later-on, denied all knowledge of it. In another experiment, no less than three students described "an unarmed person who had made no offensive remark" as whipping out a pistol and shout:

"'Stop, or I shoot!'" Apart from these singularities which can be explained by modern psychology, the average result is that in no case out of hundred-and-fifty events to be observed more than forty-one are correctly described. The most disturbing factor of pre-occupation is eliminated from all these experiments. The events enacted are all of the ordinary mechanical sort, which are not likely to touch off some pre-occupied idea. The students are eager to demonstrate their power of concentration necessary for accurate observation. Yet the reports are so very defective! While the mind is attracted by one particular event, others happen which may or may not enter the consciousness of all present on the scene. That greatly depends on the circumstances and the mental make-up of the observers.

I wonder if the enthusiasts about such "demonstrations" as the case of Shanti Devi are able to learn anything from these experimental tests of the reliability of eye-witnesses. So, I must put the point bluntly. The "verification" of the story is not reliable; it has no scientific validity; the report of the expedition to Muttra does not prove anything. The evidence of the Committee of Investigation is to be taken with a very large grain of salt. This does not imply any aspersion on the integrity and truthfulness of the gentlemen concerned. The scepticism is justified by the demonstrated fact that reports of thoroughly honest and intelligent eye-witnesses are seldom reliable. Instead of telling the whole truth, and nothing but the truth, they are never more than partially true and often altogether false. Therefore, no deduction from such evidence can be scientifically valid.

What are the "facts" of this particular case? (1) The girl is reported to have picked out of a crowd several persons and identified them as her husband, father-in-law, etc. in the former life; (2) she is reported to have directed the investigators to the house where she claims to have lived in her previous life; (3) she is reported to have given clear evidence of familiarity with the house. There are other details. But these are the salient "facts."

Now, think of the circumstances under which these "facts" were found. It was previously known at Muttra that the Committee was coming. Those picked out by the girl as her former relatives were previously informed of the story in which they subsequently figured so prominently and honourably. Yes, honourably; that's the crucial point which upsets the balance of any possible objectivity or impartiality.

The Committee was received at the station by a crowd eager to see a miracle performed. It was, therefore, in a state of great emotional excitement, which does not tolerate caution, and rules out criticism. In short, pre-occupation was the dominating factor of the mental atmosphere in which the story was "verified". How is it possible to be sure that her would be relatives did not hail the girl before she recognised them? In the given situation, such a possibility could not be excluded. Granted that precautions were taken against their actually doing so, although the reports do not say that such was the case. Even then it was impossible to control the emotions of a whole crowd. It can be reasonably presumed that immediately on their appearance on the scene, there were such exclamations as "There comes her husband or father-in-law or son!" Such exclamations would be totally involuntary, none would act with the purpose of giving the girl a tip. For the crowd, there was no doubt about the story. The idea that the girl might not recognise her relatives would not occur to anybody. So, why should any one ever think of coming to her aid? The members of the Committee, granted that they were sufficiently critical, would be naturally closely watching the girl, and consequently fail to notice the behaviour of the crowd. It would be quite natural for the credulous populace, already acquainted with the story, to speculate who might have been relatives of the girl in her previous life, in which house she might have lived, so on and so forth. Equally natural it would be for them to believe that only people distinguished for piety could have been related to such a spiritually gifted girl. Many must have aspired for that distinction; and most probably the pretenders had staked their claim publicly. Consequently, the identification can have no value as reliable evidence, unless it was assured that all the necessary precautions had been taken. The assurance is lacking. The fact is that preconditions for a scientific investigation were totally absent. The unreliability of the "verification" results from the circumstances in which the enquiry was conducted, as well as from the method adopted. The enquiry was hopelessly prejudiced by the fact that the story was publicly known at Muttra previous to the arrival of the Committee of Investigation. Having had previous information, it would be only natural for the would-be relatives to present themselves proudly and prominently at the station. One must have a very high degree of credulity to believe that, under the given circumstances, the girl's behaviour was not aided and influenced—involuntarily. The essential condition for a reliable test would be to keep the people at Muttra totally in the dark. This condition was absent. The precaution was not taken. It could not be done. The story was a public property before the formation of the Investigation Committee. The enquiry was undertaken when it was too late to conduct it under conditions which could give the most minimum guarantee for the reliability of the result.

As regards the other "facts", they were "found" similarly under the pressure of circumstances which were all too favourable for the purpose. Uncontrolled movement of the crowd, which previously knew where the house was, must have helped the girl direct the Committee. Then, in the house, who could guarantee that her actions were not forced by eager "relatives" reminding her of events of the past and indicating places, such as the sleeping room, bathing ghat, etc.? "There you used to keep your clothes, is it not?" The girl would immediately "remember." None was consciously fabricating the story. All concerned were honest—in their belief. That is exactly the point. In an atmosphere of unmitigated credulity, fictions are easily raised to the dignity of facts. Everyone present on the scene was there to see the demonstration of a truth, any doubt about which was out of the question. Granted that the Committee was the exception, it also was the victim of circumstances.

The necessary precautions not having been taken, the Committee could not possibly control the situation. It would be extremely bold for it to assert that throughout the enquiry nothing happened which could influence the girl's actions. Should the Committee make such assertion, it would be testifying against its own objectivity. Then, as far as I know, not a single member of the Committee was sceptical about the doctrine of soul and the belief in life after death. That fact alone is sufficient to disqualify the gentlemen for the undertaking.

You cannot put your own faith to test. That is psychologically impossible. The desire to test it signifies that the faith is lost. As long as you believe a thing to be true, you don't feel the necessity of verifying it. If you do so, that is with the purpose of convincing others who do not share your belief. There is sufficient reason to think that that was the purpose of the investigation. Indeed, it was not investigation, but verification. Investigation pre-supposes scepticism. There was a concrete instance of transmigration. The proposition was to verify it so as to adduce empirical evidence in support of the doctrine. Obviously, the Committee was prejudiced. Its object was not to find facts—to ascertain if what appeared in the story as facts, were really facts. Its object was to prove that the story was true.

Apart from these psychological and methodological considerations, there are other grounds for doubting that the enquiry was conducted with rigour. There are discrepancies in the details of different reports. The Committee's report avoids these details. But pressmen deal with them, and in doing so, expose that the whole affair was rather a procession to celebrate a miracle than an enquiry for checking up the veracity of assertions made by an illiterate female infant.

Thus, from the scientific point of view, the verification does not prove anything more than the tenacity of the belief in transmigration. But tenacity of belief is not the test of its truth. Obviously, the story was not put to a test. The report of the Committee is worthless as evidence. It may be respected as a declaration of faith; but as a challenge to science, it is ridiculous. The pretentious challengers have not observed the elementary rules of the game, being blissfully ignorant of the elementary principles of scientific enquiry. It is amusing that even now clever people are seriously talking of a "scientific investigation" of the question. It does not occur to these "scientific" believers that the data for this proposed investigation, namely, the report of the Committee, are scientifically worthless. Nor is it possible any longer to verify the data. It is too late to create conditions necessary for a scientific enquiry. Any scientific investigation into the question of transmigration must begin with some other story. Then, the first condition must be that the story does not spread before it is put to a rigorously controlled test.

It is rather surprising that so much excitement has been caused by this particular story. It is by no means a singular phenomenon. Tales of people remembering events of past life are frequently told in this country. They are of common currency,—even among the modern educated. But never before was there any organised verification enacted. This particular story does not seem to be any different from others; it does not offer any clearer or more conclusive evidence in support of the belief in transmigration. Even now one reads in the press reports of new phenomena. Why the scope of investigation is not extended? Why rest your case on data that cannot stand the test of criticism, and stick to a story that can no longer be subjected to a new verification? The "scientific" defenders of the doctrine of transmigration seem to be afraid of going too far afield.

Anybody with some scientific education knows that these stories are bred in the cess-pool of superstition; and that, given the atmosphere of rank credulity, in which they thrive, they do not offer any basis for a rigorous scientific investigation. Every instance of superstition cannot be regarded as upsetting the theories of science established on the solid foundation of systematic observation and rigorous experiments. A scientific enquiry must start from a plausible hypothesis; and for an enquiry, to be conducted according to the scientific method, the hypothesis must be *scientifically* plausible. The only hypothesis that could be plausibly set up on the *prima facie* evidence of the current stories, including the present one, is that memory survives death of the body. But this hypothesis is not scientifically plausible, because it is excluded by the definite knowledge that no such physical organisation as can carry memory, survives the biological event called death.

So, no enquiry would be scientific if it started from the acceptance of the stories on their face-value. It is necessary to go a step backward. The point of departure should be a critical attitude towards the stories themselves.

If somebody announces the discovery of some hitherto unknown phenomenon, such as water freezing on fire, or a man with three legs, or a stone floating on water, which, if verified, would invalidate this or that scientific theory, the first thing to be done is to ascertain the reliability of the discoverer. It is not a question of moral integrity, but of intellectual ability and the psychological state in which the discovery was made. How did he make the discovery? Where did he see the phenomenon described? Was he previously acquainted with the scientific theory which is apparently contradicted by his discovery? What was his attitude towards this theory? Did he really see the phenomenon, or is he repeating a hear-say? Is he capable of scrutinising, and correctly reporting? Is he not susceptible to hallucinations? These are some of the questions that must be satisfactorily answered before the discovery could be taken seriously, and its verification considered worthwhile.

The stories about transmigration should be subjected to similar scrutiny before they could be taken for the basis of any scientific investigation. Shanti Devi's story was not so scrutinised. One might indignantly demand: Why should we think that a little girl had fabricated such an elaborate story? But there are different ways of looking at a thing. It could be asked, not indignantly, but with more pertinence: How could a story told by a mere child, born and brought up in an atmosphere of superstition, be taken on its face-value? Is it not conceivable that the story was fabricated by others and put in her mouth? The fabrication could be done unconsciously; it would be a fabrication none the less. I do not assert that it *was* fabricated, but suggest that such stories *can* be mere fabrications. And that was a vital point to be looked into.

How did the story originate? Did the parents of the child and others who claim to have it first, previously believe in transmigration? Undoubtedly they did. That being the case, the story might have conceivably been constructed on the basis of some casual childish remark. The initial procedure, therefore, should have been a thorough cross-examination of those through whose intermediary the story reached the public. Such a procedure might have exposed the fictitious nature of the story, and obviate all the fuss made over its verification. When the data are not controlled, verification, may make facts out of a fiction. It is a fact that, whatever might have been the genesis of the story, the people initially concerned with it are firm believers in transmigration. This fact alone provides sufficient ground for the assumption that, by the time it reached the public, the story must have been highly coloured by the imagination of its original purveyors. The story as it reached the public must be very different from the original told by the girl, if she ever did really tell any at all. It is well-known how tales of the extraordinary are embellished in the course of propagation. The story might have originated in the following way, for example.

The girl heard her parents or other members of the household talk about past life in general. Devout Hindus are always talking of the *karma* in previous births. There might have been references to people remembering events of past life. In course of such conversations, someone might have playfully asked the young girl, what was she in her past life. Such conversations would be quite likely to quicken the imagination of the young girl. Whether the girl is temperamentally given to fantasy, could be easily ascertained through psychiatrical tests. This is one of the steps that should have been taken in order to dig into the origin of the story. A child with abnormal power of imagination would be quite liable to spin a fantastic yarn out of suitable materials picked up at random. Having heard fairy tales, such children would imagine seeing fairies; and would relate the "experience" so graphically and with such conviction as would assure the acceptance of the story by the credulous. If the story was the product of imagination, how is it to be explained that she mentioned one particular place and some particular persons? There is no difficulty about it. But before talking of explanation, it must be asked if there is really anything to explain. That the original story was so full of concrete details, is open to reasonable doubt. At any rate, there is no guarantee that originally the girl actually did mention the place and persons. Those details could have been interpolated by those who made the story public. But granted that the girl herself did supply the details, she could have hit upon the particular place and persons simply by hearing about them.

The only valid argument against this psychologically plausible assumption would be to establish the fact that her "relatives" at Muttra were totally unknown to the family of the girl before she revealed the secret. This all-important point was never touched. None of the investigators ever took the trouble of ascertaining this fact. Indeed, none thought it necessary to scrutinise the origin of the story. Now then, supposing that the people at Muttra had been previously known to the girl's family, she must have heard the names mentioned. Then, the name of the place, Muttra, is very suggestive of all sorts of miraculous happenings. The girl might have heard that the death of the woman at Muttra had some temporary relation with her own birth, and moreover that she died at childbirth. Furthermore, that coincidence might have suggested to some member of her family that probably the dead woman was re-born as the girl. While dwelling on the coincidence, someone might have casually or playfully asked the girl if she was the reincarnation of the Muttra woman. An affirmative answer, given by the girl, mechanically, or allured by the idea of having lived at Muttra, would start the whole story with all the details put in subsequently; and that would be done fully in good faith. The story as it eventually reached the public need not have any more substantial foundation of fact. Nevertheless, it would be a mendacious fabrication.

Once the point of departure is given by the whim or fantasy on the part of the girl, the whole story follows logically. Only, the worked out yarn was not originally told by the girl; it was put in her mouth, but those who did so were unaware that they were spreading a falsehood. They themselves honestly believed in the story. It derived its "truthfulness" from the faith on the part of the good people in the doctrine of transmigration. Given that faith, the story necessarily follows from the single affirmative syllable pronounced by the girl. If she was the dead woman, then so-and-so, of course, was her husband, so-and-so her father-in-law, her child was still living, she lived in that particular house, etc., etc. What could be more obvious? And it would only be all too easy for the girl to repeat the story as told by herself from the very beginning.

The origin of the story may be explained in still another way—even more scientific. The former explanation is psychological, the latter biological. It might sound rather far-fetched, but it is quite plausible scientifically. Of course, in this case also, the starting point is the assumption that the Muttra people were previously known to the girl's family. The assumption is permissible, because this is a point that can still be ascertained. It is not an arbitrary assumption. Let others prove that it is baseless; meanwhile, the exposure of the fictitious origin of the story is hypothetically sound.

But the investigation of this crucial point must be very rigorous. Statements of the parties concerned, however solemnly made, are not valid as evidence. Independent evidence must be sought. Meanwhile, the assumption of previous acquaintance stands. Without it, no scientific explanation is possible. Science has been challenged; so it is entitled to offer its explanation, and its explanation should not be judged by the standard set up by the alternative view of the question. Science has no plausible ground for assuming transmigration. Therefore, it must approach the story from some other angle of vision—with the object of explaining a phenomenon. If previous acquaintance is proved (and it is sure to be proved, provided that the enquiry is properly made, because it is practically certain that it was there), then, the origin of the story would be scientifically explained, and consequently, it could no longer serve as a hypothetical ground for the doctrine of transmigration. The story exposed as a fiction, the question of verification does not arise.

Here is the other hypothetical explanation—from the biological point of view. The death of the woman at Muttra might have been talked about by the girl's mother before her (girl's) birth. It appears from the reports that there was an interval between the death of the woman at Muttra and the birth of the girl. The interval has not been definitely fixed—yet another evidence for the looseness of the whole enquiry. So, it is just possible that rather the conception than the birth of the girl approximately coincided with the death of the woman. Indeed, for the convenience of the demonstrators of the doctrine of soul, this should be the case; otherwise, the transmigrators would be confronted with a tough problem.

The new organism is "born" at the moment of conception. As soon as the ovum is fertilised, a new organism comes into being. Since the doctrine of transmigration must interpret soul as the *sukshma shareer*, according to it, the embryonic body should be occupied by the transmigrating soul at the moment of conception.

But to return to the point. At the time of the conception of the girl, her mother might have thought of the recent death of the woman at Muttra. Assuming that there was acquaintance, such thought would be only too natural, particularly because the death had been at child-birth. While remembering the death of a kin or an acquaintance, she might fancy that the dead woman would be born as her prospective child. Thoughts or fancies that occupy the mind of the mother at the moment of conception are known to be deeply impressed upon the embryo, and thus inherited by the child. That is the reason for the great difference noticed in the psychology of children born of spontaneous love and of those produced by "breeding machines". That is an incidental remark. Subsequently, the mother might keep on thinking of the possibility of the dead woman's going to be reborn as her child, and talk about that fancy of hers. Such behaviour on the part of the mother would render it possible that the girl was born with a pre-conceived idea, inherited from her mother, accidentally to emerge from her sub-conscious mind, in response to some external suggestion, as "the memory of past life".

Of course, even then the story would not be so full of concrete details. These would be filled in as in the case of the alternative psychological explanation, but in a lesser degree. In this case, the nucleus of the story would spontaneously originate with the girl. On hearing Muttra mentioned, she might suddenly be conscious of the pre-conceived idea inherited from the mother, and say: "Oh, I know Muttra. I lived there," or something like it. Any such declaration or insinuation on the part of the girl would confirm the mother's premonition. She would say that she had always thought so, or felt so. The mother's original fancy would thus be immediately transformed into a "reality". And the story would be woven with all the details, just as in the case of the alternative possibility. Only, in this case, it would have a broader foundation of "fact". The girl, presumably, had never been to Muttra. So, she could have known the place, lived there, only in the past life. She says so. Therefore, she must be recollecting her past life. This assumption logically follows from the belief in transmigration.

While watching all the commotion caused by the story of Shanti Devi, who came to be such an object of reverence as to be received in audience by the Maharaja of Patiala as well as by the Mahatma,

I could not help being somewhat scornful about the intelligence of the proud and aggressive defenders of the spiritual culture of India. It could be easily seen that there were a number of serious questions which the believers in transmigration would find very difficult to answer satisfactorily.

For instance, why does not everybody remember the events of past life? Inability to answer this question convincingly, is fatal for the contention that stories of rare individuals, having the remembrance, prove transmigration. It is held that the doctrine of life after death is not a matter of belief, of primitive animistic tradition; that it can be empirically proved. Now, empirical laws are inductive generalisations. Human mortality is believed to be a law of nature, because all men die sooner or later. But in the case of transmigration, we are asked to generalise from the exception! That is a very curious idea of empirical proof. Suppose some individuals are found to have outlived the usual span of human life. The maximum claim justifiable on the strength of that discovery, provided that the data are critically checked up, would be that mortality is not a general law. On the strength of the fact that a few men have lived well over a hundred years, it would be simply ridiculous to maintain that millions and millions, actually observed to die, do no not really die! The attempt to prove transmigration empirically involves the responsibility of explaining why everyone cannot remember the past life. In the absence of this explanation, the empirical evidence is all against the doctrine. If memory is the evidence of past life, the general rule is the absence of this evidence. Unless this overwhelming evidence to the contrary is satisfactorily explained, empirical defence of transmigration is a hopeless undertaking. This explanation has never been attempted, because there is none.

The Scriptures say that to see into the past and future, one must be endowed with *divya drishti*. Since the Scriptures are infallible, the faithful must believe that those who remember events of past life are possessed of a power which accrues only to the spiritually elevated. Indeed, those rare cases are so regarded. Shanti Devi has become a minor Saint. The curious thing is that this spiritual attribute is usually claimed, in these degenerate days of modern civilisation, by ignorant people. None of the educated defenders of transmigration seem to possess the requisite spiritual refinement which would enable them to experience the truth of what they so tenaciously believe in. The *divya drishti* is denied even to our modern Swamis—those proud propagandists of Indian Spiritualism. This is rather unfortunate. Because the fact that *divya drishti* is seldom possessed by those who really possess spiritual refinements, encourages the irreverential conclusion that ignorance and the consequent superstition are the foundation of "divine" powers. And it follows logically that in the Golden Age of yore, there were so many more seers because there was so much more ignorance.

However, the doctrine of *divya drishti* is no answer to the question. It "explains" why some can see what others cannot; but it does not explain why everybody cannot remember past life. If I really had a past life, then there is no reason for supposing that I must have a special kind of spiritual power to remember it. It won't do to argue that spiritual vision is clouded by attachment, by the bondage of the body. Memory is a sign of attachment. We remember things we like or dislike. Matters of indifference are easily forgotten. So, remembrance of past life is not a token of spiritual elevation; on the contrary, it proves greater attachment than in ordinary cases. We are forced to the conclusion that, the greater the attachment to life, the clearer the spiritual vision,—a conclusion which contradicts the assumption that the common people cannot see into the past, because the effulgence of their souls is clouded by the bondage of the body. The conclusion is ruinous for spiritualism in general. Attachment, not only to this life, but also to the memory of the past, is the sign of spiritual elevation

It is impossible to rationalise irrationalism. Transmigration is an article of faith, which follows from the *ad hoc* assumption of an immortal soul. It cannot be proved. It is foolish of the storm-troopers of spiritualism to risk a battle with science on its own ground. Theirs is a hopelessly lost cause.

There are still other questions which the defenders of transmigration also cannot answer. What happens between the death and re-birth? How does the soul or the *sukshma shareer* enter the new body? At what stage of its evolution does the embryo acquire a soul? How is the choice of the next body made? These are questions which suggest themselves to anybody not blinded by faith. None of the traditional answers to them can stand scientific scrutiny. But unless these and many other equally pertinent questions are satisfactorily answered, it is simple dogmatism to assert that transmigration is a demonstrated or verifiable fact. The spiritualist jubilation over the *desired* debacle of science is dogmatism pure and simple. If transmigration and other spiritualist beliefs are found on the verifiable knowledge of objective realities, why the crusade against science? Science knows no finality. It does not claim absoluteness. It never hesitates to throw over-board established theories if they are contradicted by objective facts. If transmigration could be proved to be a fact, science would be readily convinced. Only, it refuses to take anything for granted. The spiritualists, however, do not maintain that their view of life is scientific. Their point is that science is all nonsense. The reason of the crusade against science is that it dispenses with the assumption of super-natural agencies for the explanation of life; and belief in the super-natural is the very essence of spiritualism. Disembodied spirit is a super-natural category. It is above and beyond the laws of nature. Science can explain, hypothetically at any rate, all phenomena which may defy it. And scientific hypothesis is not *ad hoc* assumption. It is logically plausible and subject to verification. If hypothetical explanations are not borne out by observation and experiments, science simply looks for other explanations. Exceptions do not definitely disprove a law; besides, very often the exceptions, on close and critical examination, are found to be not exceptions at all. On the question of transmigration, science is not perturbed by the challenge of the divine vision of ignorance and superstition. It can easily expose the fictitious nature of stories like that of Shanti Devi—can reveal their origin in the atmosphere of superstitious beliefs based on ignorance. It can defend itself against the attack of spiritualist prejudice. Fighting science, on its own ground, spiritualism suffers irreparable defeat. Attack is not the best defence always. Sometimes it is simply foolish and ruinous. That is the case with the spiritualist crusade against science.

Science and rationalism, on the other hand, can carry the fight triumphantly in the enemy's territory. Rationalist thought does not challenge spiritualism to meet science on the latter's ground. It exposes the fallacies of spiritualist thought, and rejects it on the strength of its own evidence. Let spiritualism take up the challenge represented by the above questions which rise from the belief in transmigration.

The idea of *sukshma shareer* is implicit in the doctrine of transmigration. For one thing, the idea contradicts the notion of disembodied spirit; it means that the soul is attached to the body. Secondly, if the *sukshma shareer* were a reality, conception *as a rule* should be immaculate. The physical process would be unnecessary. Why did, then, God give human beings, created after his own image, a physical structure which is a superfluity in the divine scheme of things? There should not have been males and females. The former, at any rate, could be dispensed with. All *sukshma shareers* could enter mother's womb as in the case of Saints and Avatars. What a mishap that would be for the arrogant male! And what a gloomy existence for the female—to bear the pains of motherhood without the joy of love that makes it worthwhile! I wonder how many normal females would relish being virgin mothers. However, nature, with or without the sanction of God, is happily not such a monstrosity. Conception of a new organism is a physical process which excludes the entrance of *sukshma shareer* in the embryonic gross body. Genetics and gyneacology trace step by step the whole process of evolution of a human organism from a fertilised egg. Until it leaves the mother's womb, the new organism has no direct contact with the external world. The foundation of all spiritual attributes is life. The inanimate is purely material. That which differentiates organic beings from gross material existence, namely, life, is a chemical phenomenon. An egg × a sperm cell = a new life. Where does the soul come in? If it is a disembodied spirit, whose mysterious ways are beyond the comprehension of science, well, have it that way; but then the doctrine of transmigration must be abandoned, because this cannot do without the *sukshma shareer*, which evidently is not a disembodied spirit.

There is still another fly in the spiritualist ointment. The birth (conception) of higher organisms results from the combination of two living entities—the ovum and the spermatozoa. If every living body is the seat of a soul, in other words, if the presence of soul brings about the phenomenon of life, then two souls are involved in every act of conception, the result of which is *one* body, providing home for one soul. What happens to the other? Which of the two souls contending for the prospective home, gets possession? How is the duel between souls seeking a vehicle of re-birth settled? A "scientific" explanation of the doctrine of re-birth cannot carry conviction unless these questions are *scientifically* answered; that is to say, unless it is able to give such answers as can stand the scientific test of verification under control.

The two living entities going into the formation of a new organism come from two different bodies. So, if they bring along any traces of memory, that will be only memory of the parental bodies. They, having grown in those bodies, can possibly be the seat of extraneous informations. And science has discovered that parental characteristics, mental, emotional, as well as physical, are inherited that way. But remembrance of events is not a part of that inheritance which may be traced back throughout the whole line of evolution of the species. The previous birth of any *one* human being is lived in *two places*,—the mother's as well as the father's body. It is lived as germ cells which, having no brain could not have any memory. In any case, previous birth having been lived in the sexual glands of two different bodies, the question of remembering the events of a life lived as a human being in the world at large does not arise,—within the limits of scientific knowledge.

Yet another point of fact in the process of the birth of a new organism queers the case of the "scientific" defenders of the doctrine of transmigration. Life is not interjected from outside. It is inherited from the parental bodies. This being the case, souls seeking re-birth cannot directly enter into new bodies, nor can they choose their new worldly abode. They must enter into the parental bodies, and wait for a chance to be born in a new body. But the germ cells are born in parental bodies. So, souls coming from outside, in order to live in them, are fictitious. And what is still queerer is that, supposing that souls do come from outside, each must split itself into two, one half entering the mother's body and the other half the father's. Otherwise, the duel for the possession of the new home cannot be avoided. But even then, there arises a new problem: Where is the guarantee that the corresponding halves will come together at conception?

Bodies in the state of puberty generate myriads of germ cells, only very very few of which go into the conception of new organisms. In such a situation, the probability of corresponding halves of a soul coming together again in one body, is extremely small. Here the "scientific" re-incarnationist could make a point: That previous birth is not usually remembered precisely because of the extreme rarity of cases in which it is possible. But to make this point, itself not very strong, he must throw off his whole case. He must admit that transmigration is not direct; and then he would have to prove that germ cells are seats of souls seeking re-incarnation. However, let him take care of his case, and let us follow our argument.

It is known that a vast majority of germ cells are killed off in the process of the conception of one new organism. If they are stored up through the practice of celibacy, which is more often professed than really practised, they don't go up to the brain and increase intellectual and spiritual power. They are generated in the sexual glands. There they remain, creating the natural impulse of love-life, the inhibition of which is not a way to spiritual elevation, but to hallucination caused by mental derangement. What happens to those poor souls or half-souls whose prospect of re-birth is thus annihilated? They cannot move on to newly generated germ cells to wait for another chance. Each new germ cell is a self-contained living being. It has its own soul, and therefore cannot be the home for another homeless soul. Anyhow, the chance of a new human existence in the world at large comes to only a few, and that by sheer accident. Thus, the great majority of souls are deprived of the chance of re-birth. How are those poor chaps ever to work out their *karma*?

There is still another nut to crack. The germ cells that went into the making of Shanti Devi had been there in the bodies of her parents considerably before they went into her conception. So, the soul of Shanti Devi must have been awaiting re-birth even before the woman at Muttra died. How do you connect the two? The fact (granted that it is a fact) that the woman at Muttra died just before Shanti Devi was born, disproves exactly what it is believed to prove: The dead woman's soul could not possibly re-incarnate in Shanti Devi. In order to have the ghost of a chance to do so, she should have died long before. She was still living when Shanti Devi's *sukshma shareer* was either living half and half in the bodies of her parents, or possibly reunited in her mother's womb. The way out of this difficulty would be to assert that the dead woman's soul did go directly over to the embryonic girl. This assertion, obviously, cannot be made scientifically. Besides, it involves the ruinous admission that living bodies, not only germ cells, but even a human embryo, can be without a soul. Because, had the germ cells that went into the conception of the girl or her embryonic existence been possessed of a soul, there would be no room for an interpolating soul coming straight from Muttra. Or, why should the older occupant be dislodged? The admission of the possibility of life without soul is ruinous for spiritualism, because soul was invented to explain the phenomenon of life.

Originally, soul was assumed for explaining vital phenomena. Now these can be explained without any animistic assumption. They are mechanical processes, associated with a certain physico-chemical organisation of matter. Life itself is a chemical process. Where does soul come in? It can remain as an article of blind faith—a dogma. Its position might not be so hopeless if it were possible to find some break in the process of embryonic evolution. Then it could be maintained (until science was able to fill up the gap) that the soul slips in through the rift in the mechanical process. In the absence of any such break, thanks to the self-containedness of the process of embryonic evolution, it is not possible to answer the question: How does the soul enter a new body? Inability to answer this and other associated questions leaves the spiritualists no ground whatsoever, on which they can stand in the crusade against science. When their alternative view is so palpably untenable, except as a matter of blind faith, they cannot reject the scientific view with any pretence of reasonableness. And the scientific view does not stand by the default of any alternative, but by its own merit. The spiritualist view was born of the inability to explain life in terms of the physical laws of nature. It stood in the absence of scientific knowledge. The rise of science sounded the deathknell of spiritualism. It persists as a prejudice. Tradition dies hard.

The great bulk of the Indian people still vegetate in the pre-scientific age. Hence the prevalence of the belief in miracles, in the occult, in the mysterious, in the super-natural. The thin stratum of the modern educated is so weighed down by the ballast of general ignorance that instead of being the bold bearers of the torch of scientific knowledge, they act as valiant defenders of the tradition of superstition. The pompous crusade against science is a vain effort to defend a view of life which must disappear if India is to live. The belief in transmigration fosters fatalism. Fatalism destroys initiative. Spiritualist culture has taught the Indian masses to be resigned. The spirit of revolt is unknown to them. But they must revolt; otherwise, instead of saving the world with the message of their spiritual culture, they will themselves follow other ancient peoples into oblivion. To conquer the future, the past must be shaken off. The people of India must have the conviction that man makes his own destiny. Karma, fate, transmigration, unattachment, immortal spiritual essence, Providential Ordinance—all these are ghosts out of a dead past. Let the past bury its dead, so that the people of India may live in a future brighter than the present.

CHAPTER II

PSYCHOLOGY OF THE SEER

RELIGIOSITY is not an Indian monopoly. It is more wide-spread in this country than in others because in no other civilised country, the masses are so very ignorant. Ignorance and religiosity are causally connected. Besides, what is called naturally religious temperament, is really a cultivated habit. Therefore, it may persist even in educated people capable of casting off superstitious beliefs if they want. Nor is it a matter of voluntary choice. It is a psychological phenomenon which has an interesting history. Soul, mind or personality is not a static entity. Like any other empirical reality, it also has a natural history. At any given moment, it is a sum total of past experience, the major portion of which, however, remains subconscious. Emotional or spiritual life is largely dominated by impressions and impulses buried in the subconscious mind. Hence the mystic nature of the psychic phenomena.

About the time that the story of Shanti Devi was widely advertised as the knock-out blow to the scientific disbelief in the doctrine of re-incarnation, I read in an American periodical the account of a "religious experience". In that case also, the subject was a girl of that accursed land of rank materialism. Nevertheless, the account shows how religious temperament can be cultivated, and that mystic religious experiences result from preconceived notions, being nothing more mysterious than auto-hypnosis. Such experiences are familiar phenomena in India. But they are seldom observed critically and recorded as data for psychological or psychiatrical investigation. Superstition reads in them manifestations of the super-natural; and they reinforce the religiosity, not only of the uneducated credulous, but often of the learned sceptic.

The account of the experience of the American girl is also an instance of superstition,—a cultivated habit of religiosity. But taking place in a social atmosphere different from that of India, the event was scientifically explained. Nevertheless, in the face of a scientific explanation of the phenomenon, cultivated religiosity of the girl and others of similar temperament remained unmoved in its blind faith. Therefore, I shall record the account as an appendix to the criticism of the "scientific verification" of Shanti Devi's story.

A middle aged worker of the Ford Motor Works went to a Revivalist meeting with his wife and seventeen years old daughter. The entire family belonged to the congregation. For that particular meeting, the text of the Evangelist sermon was: "I will pour out my spirit in the last days, and the young men shall prophesy and the young women shall dream dreams."

It was a winter evening, wet and cold. The parents were reluctant to take along the girl who was of weak constitution, having had three attacks of pneumonia. But the girl would not miss the meeting, because she *believed* that it was going to be a great occasion for her. She had made that enthusiastic declaration on the way to the church. Evidently, she was determined to catch the spirit of God and "dream dreams."

As soon as the preacher roused the congregation to a frenzy, usual in such Revivalist meetings, the girl rushed up to the altar and collapsed there in a heap. She was in a trance, described variously in different religions, as beatitude, *samadhi*, *dasha*, etc. Throughout the night, the whole congregation remained in the church and prayed in pious ecstasy. Finally, the unconscious girl was carried home. There, she lay in her trance day after day. The parents declined to take her to the hospital. They firmly believed that the girl was having a communion with God. The proud father exclaimed that the state of the girl was produced by the death of the sinful nature of the body.

Physicians, however, came to ascertain that the life of the girl was not in danger. On examination, no alarming symptoms were found. There was nothing wrong physiologically—the pulse was normal, reflexes satisfactory. But psychiatrists ascertained that *psychologically* (as distinct from purely physical reflex actions), the girl responded only to religious stimuli. For example, when a prayerful hymn was sung, the girl's pulsation increased; on being asked if she loved God or was with him, there was a faint smile on her otherwise entirely expressionless face. Responding to the priest's call for a fervent prayer, the girl's rigid arms shot up in a supplicating posture, while the rest of her body was irresponsive to any stimulus. She held her arms up in that posture for forty minutes, while all the conscious members of the congregation, though in ecstasy, got tired in ten minutes more or less, as any ordinary person would. That was a "miracle" which was proclaimed by the priest to be a manifestation of the super-natural power attained through the communion with God. Members of the congregation, of course, believed it to be so, and kept on praying ecstatically.

On the sixth day, having been in that state of religious coma for hundred and forty-three hours, the girl finally woke up to make the following declaration: "I seemed to be standing on a cloud with the earth below me, and I had a glimpse of the Heaven. I saw God walking towards me on a white path." She also claimed to have seen the recently dead little son of the priest, "picking flowers along the path, dressed in pure white."

Now, that was a mystic experience which, in this country, would naturally reinforce the religiosity not only of a small community, but of the entire population. The educated would read in it another irrefutable evidence of transcendental truth and realities beyond the reach of scientific investigation. Disregarding her predisposition to auto-hypnosis, superstitious mentality, debilitated physical conditions conducive to abnormal neurotic state (hysteria), in short, her general spiritual backwardness, the girl would be hailed as a seer, a *sadhu*, a Free Spirit, and could easily assume the authority of a prophetess commanding a devoted following, if she were so disposed. And the band of her disciples would be composed mostly of educated people. For, with all their traditional religiosity, the ignorant masses are not *actively* religious. Not being at all bothered by the disturbing influence of modern education, even of the most rudimentary and superficial kind, they do not feel the necessity of rationalising their faith. In modern countries, religion still thrives only in its appropriate atmosphere—among the ignorant, ill-educated, intellectually deficient, frustrated. There, religious mentality is cultivated only by those deprived of the benefit of modern education or dejected by the defeat in the fierce struggle of life. In India, religious revivalism is rampant among the modern educated middle-class intellectuals. It is mostly by their effort that religious mentality is assiduously cultivated. Therefore, mystic experiences, authentic or fraudulent, which otherwise would be lost in the vast wilderness of mass superstition, receive so much publicity whenever they happen to come to the notice of the modern intelligentsia.

The skepticism of the talented young man, who became famous as Swami Vivekanand, was swept away by the claim of an illiterate person to mystic experience which, if subjected to a psychiatrical scrutiny, would be found to be of the nature of the experience of the American girl. In both cases, the claim is personal communion with God. Science has no difficulty in conceiving the sincerity of the claim in either case, and yet show that the experience takes place on the background of superstition and, therefore, though authentic *subjectively,* is not an experience of objective truth. It is mental anaesthesia, self-hypnotism or neurotic struggle with reality.

Predisposition is the condition for mystic experience. One can be so disposed constitutionally, that is to say, possessed of a mental make-up heavily loaded with preconceived notions. Neurotic condition conducive to the struggle against reality results from inhibition of physical impulses and suppression of normal mental activities. In apparently normal persons, the predisposition for mystic experience can be produced by hetero-suggestion. Even in such cases, the predisposition is there, congenitally, so to say. It is partially overwhelmed by reason. The wavering will to believe is reinforced by the claim to actual religious experience on the part of someone who easily does the rest by means of suggestion. In authentic cases of mystic initiation, the *guru* performs this function, of placing the disciple in the state of self-hypnotism. Latent predisposition is thus awakened.

The content of religious experience, whether spontaneous or attained through laborious practices, is imaginary. Interest is focussed upon a particular image to the exclusion of other objects claiming the attention of any normal person. There is, however, nothing mystic in the experience; it is a psychological state, either produced spontaneously, or cultivated through the practice of auto-suggestion. Every psychological state can be reduced to a physiological state of the brain. Emotions are governed by internal secretions which, in their turn, are affected by emotional states. In psychological states, believed to be indications of religious experience, conscious mind lapses into a coma; the ego visits the dream-land of the subconscious. Neurotic persons fall into mental coma spontaneously. Hysteria is not usually regarded as a religious experience but as disease. *Samadhi* is mental coma attained through practice. It is a form of hysteria. Visions seen in *samadhi*, therefore, are not more divine than the dreams of the neurotic or the hallucinations of hysteria. They are images of the seer's desires. One may "see" whatever he wishes to see, provided that he has the faith necessary for the purpose. The village urchin "sees" ghosts in every bush, because he believes in all the ghost stories he has heard all his life. To have mystic visions, one needs only to disturb the normal operation of mind.

What is believed to be the attainment of a higher form of consciouness, experience of realities beyond the reach of mind, is really a psychological reaction. It is either associated with given pathological conditions as in the case of those suffering from hysteria, or produced by the canalisation of mental activity on one particular interest. In either case, there is a temporary suspension, or coma, of cerebral functions in the neo-mental, intellectual area (cerebral cortex), psychological activity, taking place then only in the paleo-mental, thalamic region (base of the brain). In other words, the more primitive, *biological*, functions of the nervous system gain predominance over what, in the absence of a more appropriate term, can be called purely mental activity. (This term is incorrect and misleading, because even the purest mental activity is expenditure of nervous energy, liberated either by external stimuli, or by the reciprocity of internal excitations).

Rational activity takes place in the neo-mental region. When that part of the brain is in a state of coma, all intelligent control and guidance of psychological activity disappear. The ego is confronted with a chaotic kaleidoscope of stored-up images shifting under the influence of nervous reflexes and mutual excitations. The operation of reason and intelligence thus suspended, there is a return to primitive (culturally as well as biologically) psychological state dominated by superstitions. Mental pictures, created by ignorance, appear as realities actually experienced. Inhibited physical impulses, suppressed desires, are sublimated into mystic images and spontaneous emotions (ecstasy).

So, in the state of beatitude, attained in religious experience, *intellectually*, man sinks to a lower spiritual level, instead of rising high above the reach of consciousness. The theory is that all activity of the conscious mind must cease as condition for the realisation of the super-human in man. Now, modern physiology shows that cessation of the conscious mental activity is produced by a cerebral coma, and therefore is coincident with the relapse into a more primitive psychological state, wherein biological reflex-action reigns supreme. In that state, man comes nearer to animal than God. The divine *shakti* released by the suppression of the conscious mind, *really*, is something which is neither divine nor mysterious. It is the life-force which man shares with all the members of the organic world; and that force is a form of physical energy. Man is spiritually superior to other living creatures because in human organism, there appear psychological phenomena which transcend biological laws holding supreme mechanistically throughout the animal and vegetable kingdoms. Before it becomes conscious, life is an *animal* force: not even human much less divine.

If *shakti* is supposed to be something other than life, then it is a mere name,—for nothing. It is a figment of imagination. Human organism does not possess any other empirical category of force than life and mind. The object of religious experience must be an empirical category; otherwise, it would be nonsense to talk of experience. While the *yoga* system differentiates life (*prana*) from the soul as well as from mind (*manas*), in the Gita, they are identified. And all the different schools of Hindu philosophy are supposed to be co-ordinated in the Gita. In the second chapter, Krishna says: "Life cannot slay; life cannot be slain." Evidently, life here stands for soul. The following line makes it still clearer; therein identical qualifications are attributed to soul. "The soul was never born; it shall never cease, etc., etc." Indeed, the idea of soul is animistic. Its existence is deduced from the vital phenomena. The divine *shakti*, therefore, is the same as life. Modern science has dispelled all mystery about the vital phenomena. *Shakti*, therefore, is animal force which in man develops the spiritual properties of reason and intelligence. Religious experience thus is the experience of the animal in man; only, thanks to pathological conditions, which must be created by practice, if not existing spontaneously, animal instincts are sublimated. Superstition becomes revealed truth; Mental images, born of ignorance, appear as supernatural realities; preconceived notions take concrete form; faith manufactures facts.

In an artificially produced subconscious state, the ego experiences the mechanistic operation of the force of life. Reflex-actions of the nervous system, which constitute the foundation of mental life, are rather biological than psychological phenomena. If conscious mind is to be regarded as the obstacle to man's experiencing the free flow of divine power in him, then, the identification of this with the animal force of unintelligent life cannot be avoided. Intelligence differentiates man from the lower animal. It is the function of conscious mind. Suppression of the conscious mind, therefore, naturally leads to a sub-human, not super-human, psychological state. Super-consciousness is a fiction which really stands for subconsciousness.

Modern biology has outgrown the recrudescence of vitalist prejudice. Neither Bergson's "*élan vital*," nor Driesch's "*entelechy*" is more of an empirical reality than the divine *shakti* of the Hindu mystic. "Scientific" vitalism ascribes purposiveness, that is, intelligence, to life which is described as a mysterious impulse. There is absolutely no empirical evidence in support of the contention. All the evidence is to the contrary. Except in higher animals with developed brains, life operates as a blind impulse. Vitalism stands on a purely logical ground, having no ontological foundation. It argues that mental activity in higher animals proves that intelligence is inherent in the vital force. Even its logic is fallacious. It leads to infinite regress, unless one stops at Panpsychism or Pantheism, neither of which doctrines can claim scientific support. Besides, consistent Pantheism is inverted materialism. When the world is reduced to one single entity which can be interchangeably called matter or spirit, the distinction between body, mind, soul, intelligence, intuition, becomes meaningless. Religious experience pre-supposes dualism. Individual soul must free itself from the bondage of body, in order to feel its union with the world-spirit. The body could not be a bondage if it did not exist. The fallacy of its logic drives spiritualist monism to an absurd position. Shankaracharya declared Maya to be real! How could the world be real and unreal at the same time? Modern absolute idealism also takes up the same position. According to Bradley, "appearances" are unreal, but they exist! There, then, exists something which is not spirit. That is a contradiction of the panpsychist or pantheist view. Vitalism is unsound even logically because it leads to this contradiction.

However, *even logically*, vitalism has no force as against evolutionary biology. The doctrine of emergence explains how consciousness appears as a "novelty." But the point is that, on the contention of vitalism itself, mind stands higher than intuition in the scale of spiritual evolution. Mind is the afflorescence of intelligence, which, the vitalists contend, is latent in life-impulse. In other words, mind is a higher form of the vital force,—the highest so far reached. May be, still higher forms will be evolved. But for the moment, we don't know anything about that. For the present, the future can be visualised only in terms of greatly developed mental activity. So, the flow of *shakti*, that is experienced while conscious mental activities are suspended, is a manifestation of life-force lower than intelligence. If there is anything divine in the life-force, it should be more manifest on a higher level of life. Intelligence is the highest manifestation of whatever spiritual or mystic power there may be hidden in life. Lapse into the subconscious state of mind, therefore, cannot be regarded as spiritual elevation.

It is a reaction,—degeneration. Religious experience (*samadhi*, etc.) is a psychological derangement,—an artificially created psycho-pathological condition. Only superstition can sublimate abnormality. The vision of the Seer or the ecstasy of the devotee is a psychiatrical phenomenon like the hallucination of the hysteric or trance of the spiritist medium.

The discovery of the cause and cure of nervous disorder throws much light on the mystery of religious experience. The famous French physician Charcot discovered the relation between hysteria and hypnosis. While treating cases of hysteria, he observed that hypnotised patients accepted without the slightest resistance any idea suggested to them. Further investigation disclosed the fact that the tendency to be influenced by suggested ideas was not produced by hypnosis. It is a characteristic symptom of hysteria. Normal persons, provided that they are so disposed, can also be hypnotised. But the mind of a hysterical patient is more open to suggestion than that of a normal person. Hypnosis is an artificially produced (as distinct from the pathological), more or less partial, coma of the conscious mind. Physiologically, mental activity is expenditure of nervous energy liberated by external stimuli or internal excitations. Under normal conditions, the mind is occupied with a variety of interests. The liberated energy is canalised in diverse directions. As expended energy is constantly replenished from the reserve put up by the cells of the nerve-tissues, there is a balance of cerebral functioning. Excessive concentration on one particular interest upsets the balance. One particular set of neurones consumes its whole reserve. Consequently, other parts of the brain are not excited internally, while excessive concentration upon one particular interest renders them incapable of feeling external stimuli. In the absence of external stimuli as well as internal excitations, energy is not liberated. There is a temporary suspension of cerebral activity, except of one particular set of neurones which itself is exhausted by using up its entire reserve of energy. Hypnosis produces such a state of partial coma, because the hypnotised person's attention is concentrated upon one particular interest. The hypnotic state is really a momentary case of hysteria artificially created. *Therefore*, even normal persons, under hypnosis, are also open to suggestions. Besides, they are "normal" only apparently. They could not be hypnotised, that is, made to develop momentary symptoms of hysteria, unless so disposed. Evidently, the disposition itself should be regarded as a symptom of the tendency to hysteria. It is a fact that only weak-minded persons can be *consciously* hypnotised, although their weakness may not always be apparent.

Hysteria, or the tendency thereto, as evidenced by suggestibility, indicates congenital deterioration of the entire nervous system including the brain.

In other words, it is a sign of a hereditary mental degeneration. This conclusion, reached by Charcot, was reinforced by the investigations of Janet who found that hysterical patients could not have more than one idea at a time. That is a sign of mental deficiency. Rational thought is not possible in the absence of the association of ideas. Of course, hysterical symptoms occur also in people who are not mentally deficient,—apparently. Even in those cases, the mind is predisposed to be obsessed with one idea. This is demonstrated by their suggestibility under hypnosis. The obsession with one idea, either "normally" or under hypnosis, releases emotion to overwhelm intelligence. Because hysterical symptoms, developed in intelligent people, who appear to be mentally normal, cannot be traced to purely physical causes,—to any specific defect of neural anatomy—therefore, they have been regarded by the religious as signs of mystic communion with God. But they can be explained psychologically as produced by the conflict of suppressed emotions; and emotions are biological phenomena. By digging into the subconscious parts of mental life, psycho-analysis reveals the causes of the phenomenon of "mystic experience." There is nothing mystic about it; that is to say, the content of such experience is not some super-natural force or transcendental truth; it is hallucination produced by the conflict of suppressed emotions.

On the other hand, *psychologically*, hysteria is auto-hypnosis. It is the result of a morbid occupation with some particular idea or emotion, to the exclusion of others. This reverse relation between hysteria and hypnosis was discovered by Liebault and Bernheim, who demonstrated that symptoms of hysteria could be produced in hypnotised persons. Charcot had ascertained that, while associated with all sorts of mental and physical disturbances, hysteria fundamentally was an emotional disorder. Modern physiology has discovered how emotion produces the physical symptoms of hysteria. Causing excessive liberation of nervous energy, too much to be used up in appropriate behaviour, strong emotions produce all sorts of physical disturbances, such as are recognised to be hysterical symptoms, and are believed to be the evidence of religious experience. Lapicque's investigations reveal emotions as purely physiological phenomena. An abnormal state of emotional excitement, which produces mental and physical symptoms of hysteria, is a state of auto-hypnosis. Because that state represents momentary suspension of normal cerebral activity, and, in cases of physical disturbance, of the entire nervous system.

Ordinarily, the mind, at any given moment, is occupied with a number of ideas checking and correcting each other. One of them may be the predominant. As a matter of fact, intense mental activity is concentration upon one single idea.

Hence the exhaustion that follows. Intense mental activity means liberation and expenditure of a large amount of nervous energy. But except in rare cases of nervous break-down, caused rather by physical debility than psychological disequilibrium, exhaustion from intense mental activity does not produce coma. Because rationalist thought always involves more than one idea, it is never concentrated upon one single interest. The very essence of rationalism is the checking and correlation of an idea in the light of others. Intense intellectual labour, apparently dominated by one idea, does not draw upon the energy reserve of any particular set of neurones; the entire brain is involved in the process. The energy expended is liberated by internal excitements. The supply being copious, it is not totally consumed even when the demand is great.

The distinctive psychological feature common to the hysterical and hypnotic states, is concentration upon one single idea to the exclusion of others, the interest being emotional, not intellectual. An idea suggested to a hypnotised person operates independent of his conscious mind, indeed, overwhelming this for a time. So long as he remains in the hypnotic state, the normal function of his cerebral mechanism is suspended. The process is more pronounced in hysteria, the cause of which psychological malady is obsession with a fixed idea. From this it is clear that psychologically hysteria is unconscious auto-hypnosis. Sometimes it is conscious. Symptoms of hysteria can be produced in hypnotised persons, because auto-hypnosis lies at the bottom of the abnormal emotional state which brings about the nervous and physiological disturbance called hysteria. In hypnosis, the idea is suggested from outside; the hysterical patient receives the suggestion from his subconscious mind. Charcot's investigations, carried on in greater detail by Janet, establish the existence of the unconscious process of auto-suggestion. Since hysterical symptoms can be produced by hypnotic suggestions, it is evident that the disease must be due to similar suggestions present in the patient's mind, without his being conscious of them.

That is the internal connection between hysteria and hypnosis. The discovery of this connection between the physical and psychological aspects of a complex pathological condition throws a flood of light on the mechanism of religious experience. It reveals the physiology of the exalted emotional state in which the mystic claims to come in sensible contact with realities beyond the reach of intelligence.

Religious experience is a case of hypnotism, the suggestion for producing the abnormal emotional state coming either from outside or arising from the subconscious mind of the subject. But, in either case, ultimately, it is self-hypnosis, that is to say, hysteria. To realise this fact, one needs only compare hysterical symptoms, produced under hypnosis, with the tokens of mystic experience.

Let us take the famous case of Ramakrishna Paramhamsa, which was so undoubtedly spontaneous as to impress many intelligent observers and convinced not a few skeptics. At the touch of any metal, his hand used to be paralysed. The cause of that remarkable phenomenon, according to himself, was his aversion to money which he regarded as the emblem of worldliness. The phenomenon *is physically explicable*. It is a reflex-action, analogous to the shutting of the eye-lid at the approach of an insect, with the only difference that it takes place on a high emotional level. And that only proves the physical nature of emotion. Like all complicated reflex-actions, the paralysis of the hand was preceded by a cerebral process—a mental activity. Money is bad; its possession involves one in worldliness, which is harmful for spiritual life; so it must be avoided. This is a whole process of logical deduction. But it takes place in the subconscious mind, completely overwhelmed by an emotional super-structure,—aversion to money. That is hypnosis, the suggestion coming from the subject's subconscious mind where the rationalist foundation of his aversion remains hidden. The aversion to money alone would not have the hypnotic effect if it resulted consciously from an intelligent conviction. An intelligent desire to avoid contact with money would not produce a physical reflex at the touch of metal wares; because intelligence would distinguish metal from money. To regard a coin as a mere piece of metal and consider this latter as identical with any other piece of matter, would be abstract thought, though from the practical point of view it would be false. When the Paramhamsa threw some coins in the river, he was engaged in rational thinking. To shrink from the contact of such things as ornaments, whose value is calculated in terms of money, would also be a behaviour determined by intelligent thought. But it is utterly irrational to believe that the mere touch of any metal is corrupting because coins are made of metal. Indeed, intelligent thought can never lead to that conclusion. The original idea about the evilness of money is totally lost—in irrational emotion. The behaviour, believed to be the conscious expression of a rational idea, is purely physical,—a reflex-action. And when it takes on an abnormal form, such as paralysis, it is a symptom of hysteria, brought about by emotional excess. It is quite natural to remove our hand automatically whenever it comes in contact with a heated object; but if the reflex-action becomes more violent, for example, if someone runs away at the sight of fire, or suffers from any physical deformity simply from the proximity of a heated object, he is regarded as behaving neither rationally, nor naturally in the physical sense. In plain language, that is a case of hysteria, and is usually given medical treatment. Similarly, the paralysis of hand at the touch of any metal vessel is a case of hysteria. It is a physical phenomenon, and should have a physical cause. To maintain that it is produced by any spiritual force, is to debase the idea of spiritual force. The essence of the idea is *qualitative* distinctiveness from the material being. This distinction precludes the possibility of any causal relation between spirit and matter, soul and body. The paralysis of the hand, even of a holy man, must have a physiological reason. As it is not a normal reflex-action, the reason evidently is pathological. There is no direct physical cause, as, for instance, a strong electric shock; so, it is clearly a matter of psycho-pathology. The reason is a morbid psychological state in which all rational thinking is overwhelmed by emotional exuberance; an hysterical symptom is produced by unconscious auto-suggestion.

Emotion is a *biological* (as distinct from psychological) function. Therefore, it affects bodily behaviour. Normal physiological processes may be disturbed by emotional exuberance. Abnormal physical behaviours take place in the absence of the intelligent control of neural activity exercised by the brain. In the absence of that control, liberated nervous energy is not evenly distributed; the excessive supply overflows the normal reaction tracks; and brings about reflex-actions without apparent cause, such as acceleration or retardation of the heart, dilatation or contraction of the peripheric blood-vessels, disturbance of gastric or other internal secretion, perspiration, profuse flow of tears, muscular contraction (in epileptic fits, cramps or partial paralysis), dilation of the pupils, so on and so forth.

That is the physiological explanation of unusual bodily behaviours believed to be indications of mystic experience. The emotional state in which they appear is not a state of spiritual elevation, but of psycho-pathological degeneration. Conscious mental activity is suspended; organic functions thrown out of balance.

Now, all these physical abnormalities can be produced in hysterical patients under hypnosis. If a patient's arm is placed in a certain posture and he is told that it is paralysed, he will not be able to move it. Charcot and his pupils made a whole series of such experiments. The suggestion that he has drunk wine would produce in the patient all the symptoms of intoxication. Given an unprejudiced approach, a spirit of scientific investigation, it is clear that in either case—of the mystic as well as of the hypnotic patient—the abnormalities must be caused by the identical reason. It is the psycho-pathological condition in which suggestibility increases in proportion as intelligence is paralysed either by disease or predisposition. It is a general conclusion of psychiatrical practice that suggestibility is greatly increased by checking the critical faculty in the patient. For this purpose, nothing is more effective than blind faith and preconceived ideas. Therefore, religious people are particularly suggestible, the suggestions being inherent in their own mental make-up. The auto-suggestion, required for producing the hypnotic state called mystic experience, is supplied either by conscious faith or unconscious predisposition. When one firmly believes that there is a world of mystery in himself, and his faith is sufficiently fortified by the bliss of ignorance, he is sure to "experience" it. All he has got to do is to take his faith seriously make it the sole interest of life.

Physiological actions can be to some extent influenced voluntarily. For example, sometimes, physical ailments can be cured or produced through suggestions. But what appears to be "volition" in such processes is not a mental force. It is stimulus to organic reaction which quickens certain kinds of glandular secretion. Even a thoroughgoing vitalist like Driesch, having investigated these "psychic" phenomena, comes to the following interesting conclusion: "It is not volition, but a kind of imagination to which must be added the firm conviction that that which is being imagined will really come about. And thereupon it happens, and that which brings it about is the vital factor operating upon the body. The things which are here at length comprehended in a scientific form, are some most ancient and others most modern; the practices of Indian yogis and of Christian Science are at the bottom the same thing in a religious form."[1] The "vital factor" is an obsession of Driesch. Modern biology has exposed the fictitious nature of this old notion refurbished by "scientific" mysticism. In the passage quoted, Driesch himself admits that unwittingly. The events ascribed to "volition" are brought about by the "vital factor"; but we are also told that, what is believed to be "volition", is really "imagination." So, "vital factor" is an imagination. Because, things equal to the same thing, are equal to one another. The finding of recent biological research is that "vital factor" is the name for the electro-chemical process which functions as the "organiser of life." However, the point here is that science has exposed the physiological essence of certain "psychic phenomena" which had previously been taken for the evidence of mystic power, whose cultivation is still believed to be the way to spiritual elevation. In other words, science has cut the ground under a venerable superstition.

Concentration upon one single interest, unless it is of intellectual nature, upsets the equilibrium of cerebral function; an hysterical state is brought about in consequence; and, hypnotised

[1] Hans Driesch, "Man and the Universe.'

by auto-suggestions coming from his belief, one sees not only desired images, but hallucinations reflecting the conflict of emotions in the subconscious self.

In the case of skeptics, converted by mystic experience, there is unconscious predisposition which is liberated as soon as the resistance of criticism gives way before the power of pure faith. This was what happened, for example, with Vivekananda. Aggressive skepticism is an interesting psychological phenomenon. It is the manifestation of the subconcious desire to find some evidence for reinforcing the flagging faith. The bold assertion of Ramakrishna that he had seen God was rather a relief (unconsciously felt) than a matter of ridicule for the young skeptic. Skepticism is not positive disbelief. It implies readiness to believe if more convincing evidence would be available. So, the assertion awakened the predisposition to believe. That immediately weakened the critical faculty of the would-be Seer. Weakening of the faculty of criticism increased his suggestibility. The silent suggestion from the Paramhamsa that the young man could have the experience of God in himself, reinforced by the auto-suggestion coming from awakened predisposition, placed him in the hypnotic state in which he found what he had been unconsciously looking for. Only, he *did not find anything* but believed to have done so. And faith moves mountains.

An authoritative opinion, very appropriate in this connection, is again pronounced by the scientific mystic Driesch. Referring to the psychological state created by imagination, reinforced by the conviction that the imaginary is real, he writes: "In this state, suggestion with reference to a particular content, that is, a faith in the reality of this content, is attained without any ground of rational nature. I pass to the other person a hetero-suggestion, that is, I tell him that this or that thing is a fact; and he internally transforms this external suggestion into an auto-suggestion, and is convinced that things are as I have said."[1]

But let us have some more facts showing that religious experience is a psycho-pathological (hysterical) phenomenon. Sometimes, hysterical patients under hypnosis behave as if they were dreaming. They would give informations about the cause of their disease which they could not while not under hypnosis. Afterwards, they would forget all about it. In the hypnotic state, hidden parts of mental life become revealed to the ego. Similarly, mystic experience is a peep into the obscure recesses of the ego; and those dark chambers of mental life are filled with distorted images representing natural impulses, suppressed consciously or unconsciously. It is also like dream. According to scriptural tradition, the divine light in man cannot be described even by those who have

[1] Hans Driesch, "Man and the Universe".

seen it. The Seer wakes up from a dream which is nothing but auto-hypnosis. The object of mystic experience cannot be described, because it is hidden in the subconscious mind. Modern psychology has dragged out the contents of those dark chambers. There is nothing spiritual in them. If honest mystics, while having their experiences, were placed under expert psycho-pathological observation, there would be much interesting revelation about the psychology of religious experience. Under hypnosis, they could be made to relate the contents of their experience without sublimating them. Of course, on waking up, they would forget what they have said under hypnosis, and would return to the religious ideas which are unconscious sublimations of suppressed desires. And there would be no dishonesty in that.

As a matter of fact, illuminating glimpses into the psychology of religious experience are afforded by the life-stories, historical or legendary, of famous Saints and mystics. The ancient Greeks associated hysteria with sex; the very term implies that. Modern psychology traces its origin to the conflict of vital impulses, one being the sexual which is repressed (mystic experience often results from the suppression of sex-impulse). Whatever may be the nature of the conflicting impulse, all Saints and Seers are well known to have suffered from the same psychological symptoms as mark the hysterical state created by the suppression of sex-impulse. They were all disturbed in their *tapasya* by the appearance of temptresses, deputed from the Court of the King of Paradise. In the language of modern psychology, they were being haunted by the images of a suppressed desire; they were suffering from hysteria. That was the psychological background of their mystic experience. Suppression of sex-impulse is believed to be the road to mystic experience. Knowing what it leads to, one cannot but regard mystic experience as the hallucination of hysteria.

Nor is it a matter of logic alone. The factual evidence is incontestable. Sublimation of sex-impulse is a typical symptom of hysteria. Women disappointed in love would devote their lives to charitable services or become nuns. Men with similar experience may depict romantic ideals of love in novels or mystic lyrics. These and many others are facts, established by modern psychological research. Even Indian religious life is rich in sublimations; because it attaches so much importance to the suppression of the sex-impulse. The practices of certain religious sects (Vaishnavas, for example) are entirely composed of such sublimations. There are others (Tantrics) who perform sexual acts, not as such, as a matter of physical necessity, but as a part of their religious practice. To make of the sex organ (lingam) the emblem of God, and to deify sex-impulse, are the most extreme forms of sublimation. To look upon every woman as one's mother, is yet another form of sublimating sex, very common in India. This form of sublimation is associated with the cult of conceiving God as the female (creative) force, or of splitting the indivisible One into male and female principles. (A curiously contradictory notion! How can the indivisible be divided? But that is a different question, which does not concern us here.)

Ramakrishna Paramhamsa was altogether obsessed with the sublimation of sex into the "mother principle." Hence his extraordinary behaviour in the proximity of women. The mother complex served the purpose of auto-suggestion to put himself in a hypnotic state. That curious behaviour was a conditioned reflex, developed unconsciously as a check upon the sex-impulse. The mere sight or touch of a woman does not awaken sex-impulse in every normal male. Then, why should the holy man fly into a trance to avoid an influence that cannot corrupt even all normal mortals? That queer behaviour was not a sign of mystic elevation, but symptom of a bad case of hysteria. In his subconscious emotional life, the conflict of impulses must have been very strong. The more fierce the conflict, the more pronounced the sublimation; otherwise, there would be insanity. That, however, does not mean that the sublimation is conscious. Indeed, in the sublimated form, the sex-impulse is totally unconscious. But it is there, —as the cause of abnormal emotional and physical appearances. To heap one's worship and devotion on a female image—on the imaginary "mother principle"—is a classic case of sublimating sex-impulse.

The religious practice of the *shaktas*—worshippers of God in a female image—includes sexual acts. To combine worship of the female principle with the suppression of sex, involves a severe emotional conflict which naturally upsets mental equilibrium, and brings about a hysterical state. All Saints and Seers, modern as well as ancient, are victims of a very complicated type of hysteria. Ramakrishna Paramhamsa, for example, lived practically always under auto-suggestions. Therefore, mystic experience was so very frequent with him.

Split-up personality is another hysterical symptom. It is often produced in spiritist mediums under hypnosis. The story of Mr. Jeckyll and Dr. Hyde is not a pure fantasy. Hysterical patients are known to develop a second personality. All recollections of a certain portion of life is lost, and the patient believes himself to be an entirely different person. Very remarkable cases of this curious phenomenon, commonly characterised as "occult," may be caused by the conversion of an emotion into morbid symptoms. A very typical case was the famous Vaishnava prophet, Chaitanya. It is said that he believed himself to be the incarnation of Radha. In trance, he enacted the part of that mythical mistress of Krishna. Undoubtedly, he imagined himself to be the beloved of God; and that imagination was the content of his mystic experience. It was, nevertheless, a hysterical symptom, produced by the suppression of sex-impulse which found the morbid emotional expression.

Nature is not easily bullied. She takes revenge which is often spiritually catastrophic, though it may be glorified by the superstitious victims and their credulous followers as token of spiritual elevation. Psychologically, the "*stribhav*," cultivated by Vaishnava mystics like their prophet, is a perverted expression of sex-impulse—of the same type as homo-sexuality. In this abnormal emotional relation, which is now regarded as a psycho-pathological phenomenon, and medically treated as such, one of the partners imagines himself to be sexually converted. The emotional abnormality originates in difficulties or frustration in finding an appropriate object of affection, which, experienced in the earlier stages of puberty, create hysterical condition. Cultivated in practice over a sufficiently long period, it becomes a physical habit with no deep psychological foundation, and thus the addict may not suffer from any mental derangement.

But sublimated in a religious devotion, its original psycho-pathological nature is accentuated.

Because, the emotion does not find even an abnormal physical expression. It develops introspectively. In course of time, the balance between emotion and intelligence is upset. The whole mental life is thrown off the gear of orderly cerebral mechanism. The devotee falls a victim of chronic hysteria which, consciously cultivated, becomes the characteristic feature of his spiritual life. That is a standing state of auto-hypnosis, and consequently abounds in mystic experience. Chaitanya, for example, would dream (in trance or ecstasy) of being in the embrace of Krishna, or caressing his feet in the classical Hindu fashion; and that image of unconscious erotic desire would be interpreted as mystic communion with God. There is nothing spiritual in such experience, which is abnormal satisfaction of a natural impulse driven underground as sinful.

The Freudian school of psycho-analysts is of the opinion that all religious symbols are expressions of suppressed emotions, chiefly of the sexual nature. Freudian "pan-sexualism," however, is rejected by other psycho-analysts; and, although Freud himself cannot be so accused, many of his followers are certainly guilty of exaggeration. Nevertheless, there is general agreement about the fundamental role of sex-impulse in emotional life. A critic of Freudian pan-sexualism writes: "We admit that further enquiry into the history of the human mind may prove that the sexual impulse has been a great factor in the development of religion and art; and there are many facts that point that way."[1] Hinduism provides a surfeit of such facts. Sex figures very prominently in Hindu symbolism. It also determines the devotional practices to a considerable extent. But the most significant fact is that even the metaphysical aspects of Hinduism are directly linked up with a prescribed attitude towards the physical impulse of sex.

Brahmacharya is the condition *sine qua non* for mystic experience. The burden of this virtue is suppression of the sex-impulse. A natural impulse cannot be killed. It is simply driven down in the subconscious mind, where it becomes harmful. It forces its way out, and influences the conscious mental life in various disguise. The control of instinctive impulses, by itself, is not a harmful practice. As a matter of fact, their *intelligent* control distinguishes man from animal. Organisation of natural impulses into a well balanced emotional life is the essence of spiritual development. That is done by reason and intelligence. Impulses, useless for, or antagonistic to, such a spiritual development, may be suppressed without any harm. If they are few and really unimportant in the scheme of intelligent life, they will remain quietly in the subconscious mind. The intellectual worker may habituate himself to light meals in order to avoid the cerebral function temporarily

[1] J. H. Van der Hoop, "Character and the Unconscious."

slowing down by the use of excessive physical energy in the digestive process. That would be a control of the impulse to eat. But the physical requirement would still be there to be satisfied by the supply of a certain irreducible amount of energy to compensate for the recurring expenditure. Only, the demand should be made through the consumption of such food as might contain greater energy in smaller quantities. That is intelligent control of a natural impulse. Similar process of control can be applied to the satisfaction of sex-impulse. One need not ruin himself psycho-physically through the practice of abstinence in order to avoid the dissipation of energy in elemental passion and lust. Indeed, sex-impulse, powerful and fundamental as it is, is automatically controlled by the operation of other emotions and intellectual occupations. The sex-impulse of all normal human beings, with a varied interest in life, is necessarily controlled, more or less. People with high intellectual occupations are less conscious of the impulse than the *Brahmachari*, who may not be anything more than a religious loafer. You cannot be really unconscious of an impulse which you are combating consciously. Control necessary for an all-round spiritual development is an automatic process. Only idle minds are swayed by elemental passions and desires. Occupied with things, not directly concerned with the ego, one ceases to be constantly conscious of the sex and other natural impulses.

However, while control (intelligent satisfaction) of basic impulses belongs to the scheme of a well-balanced mind, and minor impulses (for example, desire to go to the theatre, one particular evening, or to visit a friend, etc.) can be suppressed, the suppression of, or even the attempt to suppress, a major impulse is positively harmful.

Sex is a major impulse; indeed, the most fundamental. That is easy to see. Spiritual development presupposes life; and existence of life requires reproduction. If there is any creative principle, that is to be found in the sex-impulse. The idea that spiritual development is conditional upon the suppression of sex-impulse is, therefore, obviously absurd. The practice is as sensible as to strike at the root of the tree to make it flower and bear fruit. The satisfaction of natural impulses cannot be antagonistic to spiritual development, if the faculty naturally belongs to man. Sex being the fundamental impulse of life, its satisfaction is necessary for the physical, intellectual and emotional well-being of man. Its suppression is bound to be injurious psychologically as well as physically. In fact, it cannot be suppressed. It manifests itself in morbid symptoms, the form of which is determined by unconscious predispositions. An abnormal psychological state is created. Mystic experience is a feature of that state. It is the product of hysteria, cultivated artificially through the practice of *Brahmacharya.*

A psycho-pathological state is purposefully created to that the struggle against reality could be carried on with apparent success. Mystic experience does reveal the real nature of the self, because it is a peep into the dark chambers of the subconscious mind wherein lurk the predispositions and congenital tendencies which profoundly influence all the behaviour of life in so far as this is not guided by intelligence. But in mystic experience, realities about the nature of the self appear in distorted images. Therefore, it is rather hallucination than realisation of the truth. It is a gross superstition to seek in distorted pictures of one's inner self the guidance for conscious life. One should rather try to dig up the ugly realities behind the fascinating pictures. That would be real knowledge of the self. The knowledge of its defects, deformities, handicaps, would enable one to remove them through the exercise of intelligence, and thus liberating the self progressively from the bondage of unconscious predispositions, set it on the endless road of real spiritual elevation. Science (psychology) helping us to overcome the time-honoured superstition about the nature of self, enables us to find objective truths about ourselves instead of hallucinations. Truth is a more dependable guide to life than fantasy.

Mystic experience results from a struggle against the realities of life. In hysterical patients, the struggle is unconscious, being a product of a neurotic state. The religious man, swayed by superstition, takes up the struggle consciously. It is an instance of what Freud describes as "the escape into illness." In his case, the neurotic state is the product of deliberate practice. Once that psychological condition is created by the suppression of normal impulses and other prescribed practices, the struggle becomes unconscious. Mental life is submerged in emotional exuberance. The ruling emotions are morbid, being not the normal expression, but sublimation of natural impulses. Hypnotic dreams, hallucinations, phantasies, are the characteristic features of a morbid emotional state. Truth cannot be attained by the sacrifice of reason and realities. The light of known realities alone can illuminate the way to hidden truths.

The psychological foundation for mystic experience is predisposition which may be conscious or subconscious. An idea suggested from outside brings about an emotional state overwhelming normal cerebral function, only when it is such as awakens some predisposition. The subconscious mind is a store of predisposition acquired either through automatic experience or under the pressure of social environment. Congenital predispositions are inherited. By far the greater part of the ego is subconscious. Therefore, predispositions dominate mental life whenever emotions are not controlled by reason and intelligence. This happens in mystic experience as well as hysteria.

Religion is belief in the super-natural. Ignorance is its foundation. Metaphysical agencies are postulated by man unable to explain natural phenomena otherwise. Ignorance is the "original sin" of mankind. Therefore, no people or individual is naturally more or less religious than others. Thanks to the original sin of ignorance, religious predisposition remains deep-rooted in every human being until a very high level of spiritual development is reached. Man's spiritual life may remain befogged by the ignorance of his forefathers even when he himself has consciously cast off the "original sin."

Knowledge strengthens the highest mental faculties of rationality and intelligence. Consequently, it shakes the influence of religious prejudice. Religiosity is the badge of spiritual backwardness. Therefore, the greater the intellectual backwardness, the stronger the religious predisposition. The psychology of the intellectually backward is prone to be swayed by emotions. These are more primitive biological functions, being expressions of physical impulses. Not balanced by reason and intelligence, they conflict with each other—one trying to overwhelm the other. The result is the psycho-pathological state called hysteria. The religious, therefore, are more susceptible to hypnosis. In other words, they are more hysterically predisposed. Naturally, mystic experience is more frequent among them. That is the psychological, explanation of the religious temperament of the

Indian people. Even when the religious predisposition is hidden by a thin layer of modern education, it is still there to be awakened by some appropriate suggestion. As a matter of fact, modern education creates an emotional conflict in the religiously disposed, even when the disposition is partially suppressed by reason and knowledge. In these cases, the emotional conflict is likely to be so strong as to produce hysteria. For that reason, we have this swarm of modern *swamis* preaching mysticism as a matter of experience.

There is an internal connection between the practice of *Brahmacharya* and mystic experience. Only, the connection is not spiritual, but pathological. It is the reverse of the relation,—"sound mind in a sound body." The spiritual temperament, more correctly, religious atavism, of the educated youth is buttressed upon the traditional prejudice in favour of that pernicious practice. (It is more pernicious when only professed, but not practised; and that is generally the case.) This temperament is the psychological phenomenon of "regression" on a mass scale. An unsatisfied or abnormally or partially satisfied desire causes great mental tension and compression of physical energy, seeking emotional expression. Driven underground, so to say, it flows into subterranean channels; the result is resurgence of more elementary forms of emotion. "As the water of a dammed-up river is pressed back and flows into long abandoned channels, so the emotional tension will try to express itself in obsolete forms. Old habits, events or fantasies, which were accompanied in the past by strong emotions, will emerge once more as possible outlets for the suppressed emotion. This process is called regression, and occurs also in normal people."[1] The "normality" of people suffering from this psychological regression is only apparent. If they were psychologically quite normal, people with modern education could not be aggressively religious, and try to rationalise irrationalism, which is the very essence of religion.

To react upon the environment, is the most fundamental impulse of life. From an automatic, unconscious, physical process, the reaction gradually develops into a conscious approach with the purpose of understanding. Originally, a purely physical interaction, the relation between living beings and their environments becomes eventually differentiated into physical stimuli and psychological response thereto. The psychological content of the conscious reaction to environments is the impulse *to know*. Religion is the most elementary expression of this impulse. The essence of religion is belief in the super-natural. It satisfies the primitive man's incipient rationalist impulse to know. The significance of the impulse to know is the desire to find the

[1] J. H. Van der Hoop, "Character and the Unconscious."

cause of observed phenomena. The primitive man finds the cause in imaginary super-natural or super-human agencies. That is the foundation of religion. Imagination, reinforced by the conviction that that which is imagined really exists, assumes the complexion of "knowledge." The conviction has a pseudo-rationalist basis: Natural phenomena must be caused; they are beyond human control; ergo, their cause must be super-human. Thus, religion becomes a psychological "fixation."

This phenomenon occurs also in the process of individual spiritual development. The race being an aggregate of individuals, its psychological as well as the biological history can be traced, in broad outlines, in the development of the individual. It is an established finding of modern psychological research that a child's desire for gratification may be so deeply influenced by peculiar circumstances as to remain fixed in its earliest form of expression. That is the reason of some grown-up people behaving childishly in certain respects. Religiosity in educated people is spiritual childishness—a psychological fixation. Thanks to the "original sin" of ignorance, the satisfaction of the primitive man's desire for an understanding of phenomena, which dominate his existence, is necessarily found in faith. That peculiar mode of satisfaction puts its stamp on subsequent emotional and intellectual development. It is cast in religious mould. The twilight of primitive ignorance endures, more or less, over a long period of human development. Consequently, the forms of emotional and ideological expression are fixed by that circumstance. They persist even after the foundation of religion is undermined by scientific knowledge. Gradually, intelligence and reason, fortified by the advance of scientific knowledge, overwhelm those antiquated psychological traits. But they cannot altogether be eradicated at once. That requires a long period of time, during which new forms of expression are developed under different cultural conditions. Meanwhile, the psychological forms fixed in our spiritual childhood, and cultivated over a long period of slow adolescence, sink into the subconscious mind.

Now, mental tension caused by the suppression of a major physical impulse drives emotional energy underground. There the old forms of expression provide it with channels to flow into. What is called regression, takes place. This is an abnormal psychological phenomenon—a symptom of emotional morbidity. The recollected old forms of expression cannot be fitted into the scheme of the conscious mental life. There is an emotional conflict which overwhelms reason and intelligence. Cerebral functions are necessarily slackened, even suspended, when excessive nervous energy flows into the subterranean channels of the subconscious mind. The resurgence of the ghosts of old superstitions coincides with, indeed, is preceded by, a derangement of conscious mental activity. Regression, therefore, is a hysterical symptom. It takes place in hysterical patients. Mystic experience is a classical case of regression. It is an abnormal psychological state. It is a symptom of emotional morbidity, inasmuch as it is brought about by the practice of the suppression of the major impulses of life.

The mental tension caused by the practice of *Brahmacharya* paralyses cerebral activity; suppressed sex-impulse finds abnormal satisfaction in fixed forms of emotional expression; that is to say, in forms determined by primitive faith. The fixed forms may be conscious psychological traits as in the case of the avowedly religious; or they may lie in the subconscious mind. In the latter case, their reappearance is more plausibly regarded as mystic experience; because, then the phenomenon appears to defy rational explanation which, of course, it does not. Once intelligence and reason are consciously or unconsciously subordinated to the predisposition to find satisfaction in faith, one sees anything he wishes to see; imaginations are believed to be pictures of reality.

Faith alone, however, does not lead to mystic experience which is a rare phenomenon even in the religious atmosphere of India. It is unknown to the masses of the Indian people with all their proverbial religious temperament. They are religious because they do not know any better; that is to say, their spiritual backwardness precludes the development of higher forms of mental and emotional activity. They are fully satisfied with their spiritual childishness. Undisturbed by any departure from the blissful heaven of ignorance, their faith, though degenerated into rank superstition, does not require rationalisation. Consequently, their primitive spiritualism knows no emotional conflict, which alone produces the psycho-pathological state in which mystic experience is possible. Palpably contradictory ideas may remain side by side in the primitive mind without causing any emotional conflict. That paradox is the mark of its primitiveness. It has not yet developed the faculty of systematic thought, which is to associate ideas in a logical chain.

While for the vast bulk of the Indian people religiosity is a matter of habit, an emblem of spiritual stagnation, in the case of those with modern education, it is a psychological regression. With these latter, faith has lost spontaneity, though many of them are not conscious of that spiritual progress; and those who realise it more or less vaguely, feel distressed and endeavour to stem it. They take to prescribed religious practices which include suppression of natural impulses—physical as well as psychological—sex being looked upon as the devil of the drama. This is conscious and voluntary regression, which produces genuine emotional conflict inasmuch as the desired suppression is successfully practised. In the great majority of cases, it is not successful; but the effort itself is physically harmful. Besides, the shame of the failure creates an emotional tension which is aggravated by the anxiety to hide it. Parallel to this, there is unconscious regression. The unconscious process is more far-reaching because it is rather spontaneous than voluntary. It lies at the bottom of the religious mentality of the educated who preach the doctrine of India's spiritual mission. Modern education, provided that it is not altogether superficial, inevitably brings about some psychological change. The process is reinforced by the conditions of life in cities. Instinctive impulses tend to seek new forms of emotional, intellectual and physical expressions. Conscious resistance to the tendency would be inconsistent with reason and intelligence, which faculties are quickened by modern education. But the social atmosphere and cultural traditions are opposed to the tendency. And the mental as well as physical behaviour of man is determined by those factors. The predisposition is to resist the tendency. But it cannot be done rationally. Therefore, the resistance is offered unconsciously. Nevertheless, the result is the same,—emotional tension. Produced by an unconscious psychological conflict, the tension is not consciously felt. It finds an automatic expression in aggressive religiosity,—the endeavour to rationalise faith. The symptoms of regression are determined by the forms of spiritual expression fixed by the peculiar circumstances of the past.

The religiosity of the Indian people is a badge of spiritual infancy, a psychological fixation—as far as the masses are concerned; and with the educated, it is a symptom of regression which, being spiritual atavism, is an abnormal psychological state. When through voluntary suppression of physical instincts and emotional impulses, the abnormality is consciously cultivated, so as to develop symptoms of hysteria, mystic experience occurs in exceptional cases. The goal of divine madness is attained only by a few. Because nature provides guarantees against mass insanity.

With all their proverbial religiosity, the masses of the Indian people are no more given to the voluntary suppression of elementary human desires and impulses than any other people. While their spiritual life stagnates behind a dam of superstition, physically and emotionally they behave, on the whole, like normal bipeds. Thus, religiosity does not upset the equilibrium of their primitive psychology. Thanks to superstitious beliefs, cultivated through ages, they are totally devoid of the faculty of criticism. Therefore, they are naturally open to hypnotic suggestions particularly *en masse*. But that psychological tendency cannot affect their nervous system, the soundness of which is guaranteed by the satisfaction of all fundamental biological impulses.

In the case of the educated also, *as a rule*, nature prevails; suppression is not practised. Even those who honestly try, mostly fail. So, while religiosity is generally cultivated, and considerable psycho-physical injury results from the practices attempted for the purpose, it is only in exceptional cases that the operation of biological laws is disturbed to the extent of producing full-fledged hysteria. And searching psychopathological examination would reveal congenital predisposition in these exceptional cases. Obviously, the determining factor of predisposition is not the "spiritual nature"; because that is shared by all, and yet mystic experience is a rare phenomenon. It is either neurosis, a physical malady, or an unconscious emotional maladjustment. In any case, the Seer suffers from potential hysteria, which, aggravated by the suppression of normal impulses, develops manifestly morbid symptoms,—auto-suggestibility, hallucination, trance, double-personality, etc. Since hysterical predisposition involves physical abnormality or emotional maladjustment, it guarantees success in the practice of suppression. Complete suppression, however, is not always necessary for developing an acutely hysterical state. The attempt is enough to expose the subconscious emotional conflicts which touch off the congenital predisposition. Mystic experience is hallucination of hysteria, sublimated by superstition.

CHAPTER III

CRIME AND KARMA

IN prison, one naturally feels the absence of freedom more acutely, and for the philosophically inclined, the problem of freedom presents many interesting aspects. For example, there are prisoners—quite a lot of them—who like to be in jail. They corroborate the story that an old man committed suicide when he was released from a long term of imprisonment. There is nothing perverse in their mentality. After a long term of imprisonment, one comes out completely a misfit to the surroundings outside. For having been in jail, he is regarded as an out-caste. The struggle for life becomes for him much more difficult than previously. That being the case, it is quite natural for long-term prisoners, having nothing very particular attractive for their life outside, to get reconciled to the idea of living their entire life in jail, where at least the bare necessities of physical existence are available to all and sundry. There are other prisoners who do not mind being in jail, although they may prefer to be outside. There is still another sort who do not seem to feel the loss of freedom very acutely. In them, the spiritual callousness produced by the vicissitudes of life attains almost philosophical detachment.

Freedom, of course, is a relative concept, always and everywhere. It is limited by the circumstances of life. One is free in so far as he can control those circumstances. Man's superiority over other animals consists of the ability to exercise this control to an ever increasing extent. But the biological advantage is not realised to any appreciable degree, except in favourable social conditions. Indeed, under adverse social circumstances, man possesses much less freedom than lower animals, notwithstanding his biological superiority. This tragic fact of human existence is explained in India by the doctrine of *Karma*. This is a doctrine of social slavery.

For six years and more, I lived among about two thousand prisoners. They represented practically all the strata of Indian society, except the top most ones. The great majority of them were in jail for "crimes" committed under the pressure of adverse circumstances, in which the bulk of Indian people live. Those circumstances are not created by them. They did not choose to live in circumstances that breed crimes. Yet, there were few among those victims of adverse social conditions who did not believe that it was their fate to be in jail. The more sophisticated ones ascribed their sorrow and suffering to *karma* or the Will of God. Fatalism represents the popular conception of the law of *karma*.

According to this doctrine, everyone must enjoy the fruits of good action and be punished for the evil. The causal consequence may not be evident in the same life. The doctrine of reincarnation is the counterpart of the doctrine of *karma*. Yet, the highest ideal of Indian spiritualism is *nishkam karma*—to work unconcerned with the result. How can one be unconcerned with that which must happen to him irrespective of his will? Here is a contradiction between free will and determinism. If the law of *karma* is to be interpreted as not an expression of popular superstition, then it implies the acceptance of the principle of determinism. Providential Will is excluded from this scheme. The future of anyone is determined by his present actions, and these again have been determined by his acts in the past. Another ideal of Indian spiritualism is to regard this chain of the law of *karma* as a vicious circle, and to endeavour to find a way out to salvation. This ideal is set forth most authoritatively in the Gita.

"I do not do any work (do not act) for the result. I am neither friend to some or foe to others. I have given enough (what is necessary or what is deserved) to everyone in my creation.

Therefore, whoever knows me in this form (character) and, working not for the result, tries to be like myself, he becomes free from all bondage."

If this scriptural injunction is read together with the doctrine of *karma*, either the latter is invalidated, or an unattainable ideal is placed before the aspirant for spiritual salvation. If the law of *karma* is immutable, then, one may do good deeds without wishing the result, yet he shall necessarily enjoy the fruits of those deeds. He may go to the Heaven, and there is no freedom even in Heaven. Regarded as a determinist law, the doctrine of *karma* renders the ideal of freedom unattainable. Consequently, it militates against the higher ideal of *nishkam karma*. But there is some sense in this apparent madness. The two seemingly contradictory ideals supplement each other for forging the chain of social slavery for the masses. The doctrine of *karma* teaches everyone to be reconciled to his fate. But, on the other hand, it may make him to be more discriminating about his acts in the present and in the future. The failure to be rewarded for meritorious acts naturally discourages conformity with the established standards of virtuosity and good behaviour. Therefore a higher ideal is set up, so that one may not hanker after the reward, and behave in the prescribed manner permissible for the maintenance of the established social order.

However, there is no use trying to make the prisoners see that they are not in jail entirely for their fault, that the "crimes" for which they are punished, in most cases, are committed under the pressure of circumstances not created by themselves and beyond their control. The social background of the crimes committed by the great majority of prisoners can be easily indicated. It is so very palpable. Theft and robbery, for example, are as a rule committed under the pressure of poverty. Even the so-called "professionals" ultimately are victims of social iniquity.

Stories about the famous robber-chief Sultana are very popular among the prisoners. He has been raised to the status of a minor mythical hero. In the stories, he is depicted as a sort of Robin Hood. And it is not the heroic touches, but the social significance, of Sultana's reported exploits that appeals to popular imagination. Sultana's popularity is not confined to the prisoners. It is wide-spread among the rural population outside. Therefore, it took the police such a long time to capture him and break up his gang. I came across only one prisoner who, after his conviction for premeditated murder, had been instrumental in the capture of Sultana. He was a Brahmin landlord with some modern education and very proud of his "aristocratic" lineage.

The members of Sultana's gang, like the majority of professional high-way robbers, were recruited from the village poor. It was thanks to the sympathy and secret aid of the rural destitute that Sultana's gang defied detection for such a long time. Indian dacoits are not like the American gangsters. They are not bred out of the corrupt background of ill-gotten wealth, but are brutalised by the pure fire of poverty.

In the years covered by the observation recorded here, the number of prisoners convicted for dacoity as well as theft kept on increasing. Those were the years (1931-1936) of aggravated impoverishment of the peasantry, and pauperisation of a considerable section of the rural population owing to the catastrophic fall of agricultural prices in consequence of the world economic crisis. The income of the peasantry, already hardly enough for bare subsistance, declined by half. On the other hand, prices of manufactured articles were artificially kept up by protective tariffs, and high customs duties levied for budgetary purposes. The consequent great disparity between the prices of what they bought and what they sold further contracted the peasants' already limited capacity to buy. Pauperisation of the bulk of the peasantry contributed to the destitution of the other productive classes of the rural population, namely, the artisans. It also ruined the petty village trader.

That was a fertile field for "crimes" of violent nature. Police reports showed alarming increase of the cases of high-way robbery, dacoity and violence against the money-lenders and landlords. Popular poverty plus high taxation always give birth to banditry which is the symptom of a deep-seated social crisis. Given a highly organised State apparatus, "law and order," can be maintained by coercion and repression. But the process of social dissolution cannot be arrested. Prisons present a true picture of the situation in the country.

How does such a precarious system still stand? The answer is found in the mentality—philosophy of life, if you please—of the victims of the system. The undermined system is guaranteed not so much by political coercion as by spiritual oppression. The fundamental guarantee is provided by cultural traditions which constitute a secure foundation of the political order of repressive laws. The law of *karma* reinforces the laws of the Imperialist State. The belief in providential preordination serves as the safety-valve for the maintenance of the decayed politico-economic order of colonial exploitation, superimposed upon feudal-patriarchal social relations.

This is not the first time in the history of India that fatalism or religious prejudice on the part of the masses is assuring continued existence of a bankrupt socio-political system. In the "Golden Age" of ancient India, the masses also starved, and suffered from all sorts of misery.

There is plenty of evidence to this effect to be found in the *Mahabharat*, for instance.

Barahamihir, for example, describes a conversation between two men in a sad plight. One, voicing the spirit of revolt, naturally engendered by intolerable oppression, says: "We are suffering for the ill-doings of our King." Thereupon, the other, representing the true spirit of Indian culture, replies: "That is not true; our suffering is the fruit of our own actions in the past life." The doctrine of *karma*, the belief in the transmigration of soul, here stands revealed in its real significance. These doctrines of "spiritualist" philosophy were expounded by the *Rishis* of the old with the object of making the masses feel themselves responsible for their misery, and thus be reconciled to it. The attitude of the second man exonerates the oppressors from all responsibility, and the established social order is guaranteed against the danger of a threatening popular revolt.

This attitude still dominates the mentality of the Indian masses. Those inclined to assume an ostentatious air of religiosity devoutly demonstrate resignation to the will of God. The former are sincere in their superstition; they live blissfully in unadulterated ignorance. The latter, on the contrary, are usually consummate hypocrites. My experience with prisoners has been that the more ostentatiously religious a man, the greater a rogue he is. And in view of the fact that a Central Prison is a fairly representative replica of the entire country, I maintain that my experience provides a reliable ground for generalisation.

I picked up a stray kitten and brought it up. The way in which she was differently treated by different prisoners was a matter of interesting observation. The poor animal was invariably well treated by the "hardened criminals." Having never had the experience of kindness, these were kind. Usually, I found the "habitual criminal" to be a simple soul with spontaneous nobility hidden under a superficial callousness. I regarded the attitude towards the cat as a tangible measure of the real nature of the man. Those always invoking the *Paramatma*, or given to some other form of religious demonstration, were invariably cruel and mean. The devout Hindu would mechanically exclaim "Rama Rama" to see the cat catch a squirrel or a bird or even a mouse, which latter, by tradition, is her legitimate prey. But he would give the poor animal a good beating if he ever caught her anywhere near his food.

Theft is a greater crime than murder. That is human ethics. Murder—of the kind for which people are punished by law—is a petty matter, being violence only against individuals. Theft is a different thing. It is attack upon the very foundation of society, namely, private property. But I am a believer in the morality of theft. It represents a challenge to the immoral social order which makes "criminals" of innocent men, and then punishes them for no fault of theirs.

Habitual juvenile thieves were my particular friends. They are highly interesting subjects of psychological study. As human material, they are equally fascinating. Their philosophy of life is simply marvellous. They really enjoy life, and regard it as a great joke, though in reality it is anything but that. What to a superficial observer or a supercilious moralist appears to be criminal callousness, is really a sublime philosophical indifference to the conventional values of life. The expert pick-pocket, for example, is not so much concerned with what he picks. He is proud of his skill. He relishes the joy of plying his trade skilfully. And he looks upon his victims as so many fools. He does not steal, but pulls the leg of the society which has thrown him to the scrap-heap, and to the tender mercies of the police.

These jolly members of the brotherhood of gamins are the only honourable exceptions to the general rule of fatalist resignation, either sincere or pretended. They do not submit to but defy, fate. They have no fear for prison. Why should they? Only in prison, they are given those most minimum necessities for bare physical existence that are denied them anywhere else. Outside the jail, they are "free" to go hungry, to get wet in the rains, to shiver in the winter and to be eventually picked up by the police. Asked why he comes back to jail always after his release, one of my particular friends gave the disarming and thoroughly convincing answer: Freedom does not fill my stomach, and the police do not let me alone." And he added with a charming candour that he was glad that they did not. Two full meals a day, a dry place to sleep, blankets in the cold season—these are preferable to "freedom" to starve, for a change. And, fortunately, for those who need it so badly the change is not left to the chance. The police are there to look after it that they get it regularly. The life of these boys, being so very well determined by the very nature of their existence as the ugly symptom of a rotten social system they know neither God, nor *karma,* nor even fate. More correctly, fate is such a stark reality for them, that they have no occasion to speculate about it.

CHAPTER IV

WHY MEN ARE HANGED

THE following few cases, selected at random from among the inmates of five prisons, clearly reveal the social background on which "crimes" are committed in India. *Everywhere* crime is a social phenomenon. Crimes are committed not against society by individuals who are punished as criminals. They are crimes *of the society*. The victims of social injustice are punished by the rules of society. Since crime is the ugly bastard of social injustice, its forms are largely determined by the specific social structure of the country. The cases I have selected are typically Indian. They could be committed only in the peculiar social atmosphere of this country. Placed under different social environments, these men and women would not be hanged or otherwise punished. The motive of crimes committed by them springs from the established relations and traditional taboos of Indian society. It is to be traced not to human nature, but to Indian nature, which is the product of the Indian mode of living, and is dominated by the Indian mode of thought.

I

It was a bright lad, hardly above twenty, born of a high-caste family of small landowners. His parents had done the unprecedented thing of sending him to study in a college. In quest of the benefit of higher education, he had to go far away from his native village in the hill regions of Kumaun to some large town. A couple of years in circumstances so very different from those of his native place, had unsettled the emotional life of the impressionable youth. He had returned home, not so much with new ideas as with awakened impulses and desires which, in the atmosphere of his native village, could not but lead to disaster. I saw him under the sentence of death, awaiting execution resignedly, if not quite cheerfully. For, as he declared himself with tragic frankness, was he not going to atone for his sin? It was operation of the inscrutable will of God—of the inexorable law of *karma*. That was his own explanation.

What was the crime for which he was hanged? He had fallen in love; and, having lived his adolescence in the unsettling atmosphere of new-fangled ideas, was partially free from the inhibition against the physical basis of the emotion. Of course, sex-impulse was still a matter of shame to him, as with the vast majority of Indian youths, notwithstanding their modernism in other respects. And, unfortunately, thanks to the social atmosphere of the village, it drove him towards something still more shameful. He committed, or nearly did, incest.

The difference is of little significance, and need not be factually established. The point is that he was driven that way, because, under the given conditions, there was no other channel open for his emotions to flow and his natural impulse to be satisfied. Instead of stopping to think how such unnatural, immoral and sinful development could take place, the village, of course, was not only scandalised, but enranged. Fierce conflict of two powerful emotions, love and shame, must have driven the boy to a state of hysteria verging on madness. The stark impossibility of ever being able to escape merciless social persecution evidently made him desperate. Utterly helpless and hopeless, he revolted, and killed his beloved—to spare her a whole life of shame and resulting destitution. He did not kill himself, because he wished to atone for his sin. Having sent his beloved where no cruelties of this world could reach her, he delivered himself up to the police, and eventually walked to the gallows, not repentent, but resigned—the hero of a heart-breaking tragedy.

He had fallen in love with his own sister, recently widowed in the bloom of youth. Oh, what a horrible, disgusting affair! The moral sense of people with any decency cannot but be outraged by such lewdness. But thank God, such loathsome, unnatural incidents are very rare. Indian culture and social codes exclude the possibility of the occurrence of such revolting moral depravity and sexual looseness. But don't be carried away by moral indignation. Ponder over the facts of the case, and the deplorable situation, in which such tragedies happen, may appear to you in an altogether new light.

The youth was sent to the town for education. Evidently, he was expected to learn something that could not be learned in the village. It was not his fault that he had lived up to the expectation. One cannot breathe in the atmos phere of modern education, however defective, half and half, so to say: Take in so much as necessary to be a clerk or a minor official or a lawyer or a medical practitioner, and shun the rest as a source of corruption. In any case, a town is entirely different from a small village. For good or evil, (I believe it is for the good), the mentality of the body of Indain students in higher educational institutions is in a flux. The belief in traditional values, moral as well as social, is greatly shaken, though far from being overcome as yet. New ideas and strange ideals are agitating the youth—of both the sexes. No impressionable adolescent could live in such an atmosphere without being carried away by the spirit of the time, if not as yet positively of revolt. The inevitable disturbance of ideas and emotions is bound to be all the greater in cases of those who come to this atmosphere of partial freedom and incipient revolt suddenly, having lived until then in the placid backwaters of the remote village. The hero of our tragedy was one of such cases. In the town, he saw things not to be seen in his native place. He heard voices that had not reached him before. He allowed himself to be influenced by the strange sights he saw and disturbing voices he heard. Why should he not? Had he not been sent to the city to learn precisely what could not be learned in his village? Thus, after two years, he returned home, not indeed fully possessed of modern education but with some new notions about life.

Naturally, the young man would find the life in the village rather dull. There was no excitement, no romance, no movement, not even expectation. All channels for venting his newly awakened emotions closed, he took to brooding. The condition of his sister, a common-place of callous disregard, provided him with food for thought. A part of his education in the town had been to be conscious of the sex-impulse. That being a lesson taught by nature, he would have learned it anywhere. Two years in the atmosphere of semi-modernism had only encouraged him to be honest, at least to himself, if not to others. Most probably, he still regarded his willing response to the irresistible call of nature as a shameful thing to do. Yet, presumably, his modern education had gone to the extent of making him admit to himself, perhaps with great reluctance, that he could not deny feeling the impulse. That much education would make him realise the difficulty of the tragic life to which his sister had just been condemned for no fault of hers.

So damn ably callous is our social cruelty tha fathers well past middle-age keep on diligentlyt at the marital duty totally oblivious that widowed daughters or daughters-in-law, still in the bloom of youth, are made also of flesh and blood, and consequently might find it extremely difficult to live up to the ideal of virtuous renunciation forcibly set to them. The traditional taboo on the frank admission of sex-impulse makes the unthinking herd unconscious of the cruelty they callously commit. Partially awakened from the moral coma, the young man saw that his sister was also young and as such could not help feeling the exigencies of youth, although she was compelled to pretend that she did not.

In such a tragic situation, it would be but natural for a brother, no longer quite an unquestioning conformist with all the stupidities of village life, to come to the aid of the distressed sister. In the town, he had seen young women going about more freely than in the village. He had seen them taking interest in things, forbidden, or unknown to, or beyond the reach of, the rural women-folk. He had seen that life, less restricted by old social customs and prejudices, made women more cheerful. He thought that some occupation more congenial and distractive than domestic drudgery might lift a little the dreadful gloom hanging over his sister's life. He began teaching her to read and write; and, while imparting the merest rudiments of it, talked more about the benefit of modern education. The family, of course, did not look upon the procedure very kindly. According to tradition, the girl, in her state, should not take interest in anything of this world; she should devote herself to the sacred memory of the dead husband and to the service of the living relatives. However, not much attention was paid to the matter in the beginning. Let the unfortunate creature be consoled a bit. The whole life was still ahead of her,—to work out her *karma*. Besides, thanks to his "high" education, and particularly his sojourn in the town, the young man had acquired a certain amount of authority even in the eyes of the elders.

Before long, the next step was taken. The girl was persuaded to accompany her brother out of the house; the two were often seen walking rather happily in the neighbouring fields and woods. What an unpardonable thing on the part of a widow! To be happy! When she had just been condemned by Providence to a life of mourning, sorrow, privation and service! But the fat was in the fire. The young man had told the girl everything he had seen, heard and learned in the town. She avidly drank of the fascinating picture so very very different from the drab life she had lived, and the dreary and dreadful life that still had to be lived. She dreamed of freedom —of a life with some meaning. And to the youth, life is identical with love.

They planned to go to some city together—to see more, to learn more and—to live freely as they could never do in the village. They themselves did not know what they exactly meant or wanted. Perhaps they did not mean anything which would make them more than brother and sister. Perhaps they did. In any case, it would be quite natural for them to be sexually attracted, that is to say, fall in love. The young man had returned home with awakened sex-impulse, and no longer quite ashamed of admitting the desire to himself. As a matter of fact, very few normal beings are. The pretension is hypocritical. Back to the village, he might have relapsed into the habit of inhibition or taken to something much worse—self-abuse, sodomy or clandestine sex-intercourse under sordid conditions. Negative morality, imposed by our social conditions, frequently encourages these practices which are worse than immoral, because they are physically harmful, aesthetically revolting, and corrupt the emotional life of the youth, making them vulgar, often vicious, and generally hypocritical. Fortunately as well as unfortunately, there was his sister who, thanks to her own misfortune, provided him with a channel for his youthful emotion to flow normally. It seems that she responded. The call of nature was irresistible.[1]

One hears so much about human nature that never changes. The fact, however, is that the desire to love and be loved is the only constant of human nature. And the foundation of this essence of what is called human nature is biological. You can disassociate love from sex just as much as you quench thirst without drink, or satisfy hunger without food. Love, of course, is not limited to sex relations. But primarily, it is nothing more sublime or mysterious than emotional reflex of the urge of an organism to reproduce itself. Emotions, however noble or

[1] For the benefit of those who may be morally scandalised, even if sympathetically inclined towards the unhappy victims of social cruelty, it may be mentioned that incest is really not so much revolting to human nature as it is generally believed to be. The famous anthropologist Sir J. G. Frazer is of the opinion that incest was declared to be such a dreadful sin and so strict taboos were placed for its prevention, precisely because of the general tendency towards it. He came to the conclusion by exhaustive studies of human behaviour under all conditions.

sublimated, are physical urge, in the last analysis. Therefore, normally, when psychological inhibitions and social taboos are absent or swept away by extraordinary combination of circumstances, love breaks the bonds of sublimated hypocrisy, and not always successful self-deception.

Youth cannot help attracting and being attracted by youth. Segregation succeeds at the cost of sincerity; sexual chastity, bought at such a high price, is spurious. As a matter of fact, segregation breeds exactly those vices it is meant to check. It encourages promiscuity, which often amounts to incest. Forced widowhood plus segregation puts premium on incest. This is not an inferential assertion. It is a matter of fact. Have the courage to look ugly reality in the face, lift the veil of cant that carefully covers the skeleton in the cup-board, stop whitewashing lies, dig for the truth, and you will be shocked to find ample verification of this statement. I have actually heard people, not altogether illiterate, and uncultured, justify the practice committed, of course clandestinely, in their own household with the argument that the scandal had better be kept at home. Incest, after all, is not so unnatural as it is generally held to be. The priesthood of Gaya, for example, is a strictly inbreeding community. As a matter of fact, incest is a natural inclination. No less an authority than the great anthropologist, Sir James

G. Frazer, is of the opinion that precisely because of the strong general inclination towards incest, it is so severely punished by communities with greater possibility of such intercourse. Love being a law of nature, brothers and sisters cannot help being lovers when, in the critical period of adolescence, they are thrown upon themselves either by the clannishness of primitive people, or by the spiritual taboos of civilised society. That is exactly what happened to set the stage for the tragedy here recorded.

The young man went to the gallows because he had done the only honourable thing that he could possibly do under the circumstances, utterly beyond his control, if the girl he loved, either as sister or sweet-heart, was not to be driven out into a callous world, which would accommodate the derelict only as a forced vendor of herself. How painful facts of immorality result from the sanctimonious fictions of moral indignation! The unfortunate girl could not live in the village. She was sure to be thrown out by her parents, who would not dare protect her, even if they cared to. She would not be given another chance even to make good her widowhood. A widow, caught in the act of violating her obligatory virtue, or even suspected publicly of doing so, can expect no mercy from the draconian codes of our society. The facade of purity must not be soiled. What is punished, is not immorality, but revolt against, or disregard for, the established conventions. Even the lack of sufficient fear for this is an unpardonable offence. The poor girl could not possibly remain in the village. Where could she go? How could she live? These questions did not bother her persecutors,—those passionate admirers of the sublime ideal of widowhood, those stern defenders of a social system that breeds hypocrisy in the name of morality. But the moral society would not guarantee her the barest subsistence, unless she was ready to pay for that privilege. If she sold herself, she could even have a life of luxury as long as she would have her youth to sell. In orthodox Hindu society, the woman can live only by selling herself either in wedlock or in the open market. Is it moral to drive a girl to prostitution, the only alternative to suicide open to her under the given conditions, simply because she failed to obey social laws that violate the law of nature?

The young man passionately believed that it was not. The murder of the girl was a desperate protest against the tyranny of a callous society. That was the crime for which he was hanged.

II

The next case can be regarded as the sequel —not directly, but taxonomically. If the murder was not committed as it was done in the case just described, it would most probably be committed in a different form, at a latter stage of the social tragedy. If the denouement is postponed, it takes place later ; the tragedy ends with a scene in which the major role is differently played by a different person.

This is the case of a young widow, convicted, not indeed to death, but to a long term of imprisonment, for having strangled her child at birth. The shocking act was so obviously involuntary, done in utter despair, that the judge, while sentencing the accused according to the letter of the law, recommended her for mercy. I do not know if the mercy was granted.

She came from an ordinary peasant family. After the death of her husband, she had been living with his people as is customary in such cases. Only in well-to-do families, young widows can have the poor consolation of returning to their parents instead of being forced to live with strange people with whom they have no longer any connection, and who usually treat them as beasts of burden, not even kindly, they being regarded as embodiments of misfortune. For some tortuous reason, young widows are held responsible for the untimely death of their husbands, and consequently, also for the material loss suffered by the families concerned. Death of able-bodied male youths amounts to considerable material loss for joint families living mostly on the proceeds of the labour of their own members. Widows are held responsible for the calamity, because it was their fate to be widowed. The families, into which they are married, not by their choice (that little fact is not taken into consideration), are adversely affected by their bad luck, which is the result of their own sins. The belief is that a particular youth would not have died, thereby causing material loss to his parents, had he not been married to a girl destined to early widowhood. It would be equally reasonable to believe that the young man died because it was his fate to die prematurely, and the girl suffers for no fault of hers. However, that is not done; the blame is laid at the door of the widow, and she must pay the penalty. Indeed, sometimes young widows are treated as if they had actually killed their husbands. They are often called witches who have devoured their unlucky husbands.

Among the working peasants, women cannot be segregated. Purdah, where it is observed, is rather a matter of formality. There, sex morality is not meticulously dressed up, as among the upper classes, in rigid social conventions and high sounding phrases. It stands out naked as jealousy and possessiveness. The wife belongs to the husband; social morality demands that his proprietorship should be respected by others who are entitled to reciprocal consideration for their rights. Jealousy is outraged sense of ownership. It is righteous indignation against trespass. The chastity of women in general, that is, their sexual subordination to particular men they are respectively allotted to, is protected as the collective possession of the male. Such a rustic sense of sex morality, in the necessary absence of segregation, except as a mere formality, is bound to make room for what is called illicit intercourse. Conventions are there; but not cant, which is immensely more powerful, since it influences the victims of sex subjugation psychologically, thus transforming the coerced into voluntary, convinced, even enthusiastic conformists. Moral cant is absent among the lower classes, because they lack the cultural attainment which is the condition for that virtue.

Such was the social and moral atmosphere in which this particular woman lived. As long as she gave her keepers sufficient satisfaction as a beast of burden, they were not any too watchful about what she might do as a woman. Incest being a taboo by common consent—unless performed *sub rosa*, as is done not infrequently, only without love, young widows are possessed rather as beasts of burden than as women. Indeed, with the backward peasant masses, women generally are beasts of burden. In the first place, performing the sexual function only incidentally. So, the woman of this story, like others in her position, must have had a certain amount of freedom regarding the satisfaction of sex-impulse.

Those of her kind are not handicapped by psychological inhibitions cultivated among the upper classes. Nor are restrictions on movement so very strictly observed that they could not be broken through on the sly. Young widows perform labour for the merest keep. Consideration for the value of their labour does not permit their keepers to be too exacting about their conduct when they are not gainfully employed. One does not turn out such a valuable asset simply because she happens to meet a man in the barn. This particular woman was given that latitude, willy-nilly, if not quite consciously.

In course of time, she became pregnant. Even that fact, when it came to be known to her keepers, did not provoke moral indignation sufficiently strong to turn her out. For the good horse-sense of the peasant, unencumbered by moral cant, it would be silly to lose the labour power of a strong young woman on such an account as could be easily settled. She was ordered to keep under cover until the child was born, then get rid of it promptly by strangulation, and dispense with the remains neatly according to instruction. The alternative to obeying the order was expulsion from the only refuge in the hostile world, public disgrace, and a future too dreadful to visualise. If the helpless woman was actually asked to choose between the alternatives is not known. However, she had to choose. There was no other way of escape. The man she met in the barn, of course, would take no responsibility. Nor had she any claim on him. She chose to kill her child, failed to cover up the deed, was charged of an heinous offence, found guilty, and condemned. The whole tragic process was formally proper and legally justifiable, if not quite just. But perverse indeed is the sense of justice and morality which fails to sympathise with the unfortunate creature as a helpless victim of an immoral, unjust, cruel, stupid society. It was so obviously a case of crime committed under compulsion by an utterly helpless and hopeless victim of social conditions, that even the legalistic bias of the judge was overwhelmed by pity.

In her statement during the trial, the woman declared that she had killed the child, not for the sake of herself, but to spare the child the pain of dying of starvation eventually. Had she not acted as she did reluctantly, she would be thrown out in a hostile world, with absolutely no means of subsistence except charity from the merciful, which could not be expected by a branded sinner. Despair overwhelmed maternal instinct. Or, was not the crime really suicide of maternal instinct at bay? At any rate, the punished "criminal" was certainly not responsible for the crime. The responsibility cannot be laid at the door of any single individual or of a few individuals. It was a collective responsibility, as the responsibility for practically all crimes is.

III

An incident reported in the press about the same time throws light on the fact (though few saw the light) that similar sort of crime is committed also by higher-class people, supposed to be morally more elevated, but who usually escape public detection and legal punishment thanks to their advantageous social position. A smashed-up carcass of a new-born babe, well wrapped-up in good clothes, was found under a railway bridge over a big river. Evidently, it had been thrown out by some one from a running train the night before. Now, who could have done that? Certainly no passenger travelling in an ordinary third or inter-class carriage. He would be caught forthwith red-handed, because he would never be alone. To do the deed neatly, without the danger of detection, the perpetrator must travel second or perhaps first class; and since even then he could not be sure of being alone, very probably he travelled in a reserved compartment. In short, all the facts of the case went to prove clearly that people concerned with the incident were well-to-do.

There could be no question about the body being that of a murdered child. Circumstantial evidence all pointed to that conclusion. From the sort of clothes, in which the body was wrapped, it could be inferred that the child did not belong to the lower class of people. On the other hand, the kind of people, to which the dead child evidently belonged, would not leave the body in the condition it was found, had the child died naturally. In that case, it would be properly cremated. Besides, it was not necessary to depend only on circumstantial evidence. Direct evidence could be easily found in an autopsy of the body. Even a careful examination for extraordinary symptoms would do. Unwanted new-born babes are usually disposed of by strangulation which leaves an unmistakable trace.

Evidently the child was illegitimate; otherwise, it would not have been disposed of in that way. The legal crime of murder must have been preceded by the moral delinquency of "illicit" sexual intercourse, involving either a widow or an unmarried girl. Most probably, in this case, the murder was not committed by the mother. Therefore, there was no extenuating circumstance as in the case of the punished woman in the previous case. The crime was all the more heinous because it was wilful. There was no material compulsion. The motive was the lack of courage to stand up openly against stupid social conventions on the part of those who surreptitiously indulge in practices they themselves condemn as immoral when others are concerned. That is a sordid motive. Yet, the criminals escape detection, disgrace and legal punishment, because they were rich, possessed of the means to cover up their immoral conduct and criminal act. The police could have easily run down the person who had thrown the body out of the train, if they followed up the clue which was plentiful and clear. But there was some hidden force to arrest their zeal. When the rich are involved, the wheels of the administration of criminal law, so merciless otherwise, turn slowly, and leave the criminals alone if that could be possibly managed. The verdict of society also is as accommodating and condoning in the case of upper-class misdemeanants, as it is harsh and merciless when dealing with the helpless.

IV

There was an old man, well above sixty, sentenced to death for killing his young wife. On appeal, the sentence was reduced to transportation for life. His advanced age was taken into consideration. I wonder how he liked the change. Since the shortest term that a lifer must serve is ten years, there was little chance of the man serving out his sentence. It is better to die at once than undergo the process over a period of a number of years. Yet, I have not met one single prisoner who would not prefer a life sentence to the capital. Curious!

A prisoner's life is mortgaged. He cannot dispense with it voluntarily. One should not be allowed to cheat the law. Justice is a vindictive Goddess; and therefore she is so very unjust. Violation of the sacred right of property is the greatest of crimes. Criminal law is an instrument for the defence of possession. Penology is not corrective, but coercive. One is punished for having violated or threatened the right of property of others, directly or indirectly. Yet, the essence of his punishment is the deprivation of the very same right! Is it not illogical and unjust? There would be some justification if he was deprived only of those things that he had acquired from others. That would be mediaeval justice; but at least justifiable on the principle of might is right, which still remains the foundation of civilised law with all its legalistic hypocrisy. But to deprive one of the only birth-right—of that which is a gift of nature—that is really a crass contradiction of the fundamental principle of criminal justice. My life belongs to me. Only my parents can claim any right on it on the ground that it is their creation. But even they were merely instrumental in the creation of my life, which was an act of nature. So, according to the principle of criminal justice, which includes penology, one's right on one's own life is inalienable. He should have unlimited freedom as regards it. But a prisoner, whatever may be his formal sentence is deprived of this right. Logic as well as morality demand that each victim of criminal justice be given the option of killing himself instead of undergoing the punishment inflicted on him. The option could be easily given, because very few would exercise it. Fear of death contributes so very considerably to the slavery of man.

However, let me record the story of the old man who killed his wife. When he married, (I do not remember whether it was for the third or the fourth time), he was not yet sixty. So, his action was not altogether unjustifiable. But he made the mistake of overlooking the fact that the justification would disappear by the time his bride would be a woman. She was hardly ten at the time of their marriage. When she was twenty, the man was well over sixty. He naturally liked having a young wife—to cook his food and otherwise look after him. But a healthy young woman of twenty wants something more out of marital relations. The cause of the eventually committed crime and its punishment originated in that divergence of the natural needs of two people united inseparably.

The old man could not be held responsible for marrying a girl of ten. He was a simple villager; where he could get a grown-up bride in the village? Presumably, he would have preferred a mature woman for the consideration of domestic management, if not for more intimate reason.

However, he had to take what was available, like everybody, and wait. Only, meanwhile, he grew old, less capable of performing his marital duty to the satisfaction of the other party concerned, than of demanding his due which was rather of servile than marital nature.

The result was that the young wife began to carry on with some village beau. Conventionally, she deviated from the path of virtue,—sexual faithfulness to the husband. If the husband was, for some reason or other, in this particular case, for advanced age, not capable of retaining the faithfulness on the part of the wife, the latter should remain faithful to the ideal of wifely chastity. Our social law-givers wanted to improve upon nature and rectify a mistake of the Maker. Sex-impulse is natural, and demands satisfaction. If it is a sinful desire, why did not God exclude it from the scheme of his creation? It is sheer sophistry to argue that he created human beings with the sinful desire in order to give them a chance to be virtuous. If God is good, everything he created must also be good. The moralists are the most immoral and the religious are the most irreligious people in the world. To try to improve on the scheme of God's creation is to insult God. Instead of abandoning their perverse, irreligious notions of morality and stupid social taboos, the orthodox persecute those who obey the will of God and act according to the providentially ordained laws of nature. The young wife of the old man could not obey the man-made law and natural law at the same time. She obeyed God instead of man. Hence she was sinful in the eyes of society which places human wisdom above divine dispensation, while pretending to be guided by this.

The young woman seems to have been particularly spirited, possessed of admirable courage and frankness, the appreciation and practice of which would improve the moral tone of our social life. The old man told that he did not intend to kill her. He wanted to persuade her to give up her evil ways. But he was uncontrollably enraged by her insolent retort to his rebuke. The scandalising answer was that, being herself young, she wanted a young man; and as she could not leave her old husband and marry again, she was compelled to go the way she was going. She further said that she was doing her duty to the husband—taking care of his household, looking after his physical welfare, etc.; that it was he who was failing in his marital duty, and therefore it was absurd for him to accuse her of infidelity. He should not be a dog in the manger; he should not object to her seeking elsewhere, in the only way possible under the given social conditions, what he could not give her.

That was an unheard of insult. It made the old man see red; and he killed her. Even after-conviction, with at least ten years of rigorous imprisonment staring him grimly in the face, the old man was boiling with moral indignation. But one could easily see that it was not so much his moral sense as the sense of property that was outraged. The woman belonged to him. She had no business to give herself to a loafer and have the cheek to justify her action in such a brazen manner. She should have at least pretended faithfulness, as the token of formal admission of the husband's right of ownership, and then carry on with the young man clandestinely, if she were so incorrigibly perverse as to desire something more than keeping house and sharing a barren bed with an old man.

The moralists do not see that, to place a woman in such a position and keep her there under the threat of social persecution, amounts to denying her the right of motherhood,—a right given by nature to every woman. Can anything be more immoral and sinful than depriving a woman of the delight of motherhood? No moralist would brand motherhood and the desire to have children as sinful. Then, how could they reasonably maintain their stupid attitude towards sex relations? There is no other way to motherhood. Only prophets and *avatars* are immaculately conceived. If motherhood is divine, the way to it cannot be sinful. When a natural impulse is placed under stupid taboos, it is bound to seek satisfaction in tortuous channels. The notion of morality that runs counter to the laws of nature is immoral. It leads to corruption and crime.

V

A middle-aged woman was sentenced to a long term of imprisonment on the charge of complicity in a clandestine act involving two deaths. The charge was not murder. A grown-up son of hers was also convicted as an accomplice.

The family concerned belonged to the urban trading class,—fairly well-to-do. The dead body of a new-born child was found in a lane by the house in which the family lived. Police enquiry led to the disclosure of the fact that, in the night preceding the discovery of the child's body, a young widowed daughter of the family had died of "fever". Finally, the story turned out to be that it was a case of abortion, the mother as well as the child having died in the process. Except the sick daughter, the mother was alone present in the house that night. So, she was arrested. The son was accused of complicity or connivance, on the strength of the fact that the day before the death of his sister he had called in a doctor to treat her for fever. The doctor had found the girl lying all covered up in a feverish condition. He also was suspected of connivance, indeed of assisting at the abortion, if not directly on the spot, at least through the administration of some suitable drug; but the case could not be brought home to him.

Whoever might have been involved in the tragedy, and whatever might have been the relative gravity of the parts played by them respectively, the salient facts of the case are these: A young widow belonging to a respectable family, despite the purdah in which she lived, had clandestine sex intercourse, if not a love affair. The result was pregnancy. She kept her state a secret as long as possible. Eventually, the mother came to know about the misfortune that had overtaken the family thanks to automatically observed social customs. What was to be done about it? The good name of the family had to be saved. The purpose would not be served by turning out the misbehaving daughter. Besides, the "misbehaviour" might have all along been a skeleton in the family cup-board. In cities, the movements of purdah women are strictly restricted. Any misbehaviour, therefore, must take place under the very nose of the defenders of forced chastity and obligatory faithfulness. Therefore, public disclosure always affects the name of the family concerned, which is suspected of moral laxity, at least to the extent of conniving with, if not actually encouraging, the misbehaviour on the part of individual members. Thus, the mother of the erring daughter was in a delicate position, and naturally sought the only way out, namely, abortion—an act neither legally permissible, nor morally tolerated.

Moral objection to abortion, practised under compulsion, that is, by women who would be socially persecuted if they owned up their motherhood, is a curious attitude. In the opinion of the moralist, as well as in the eyes of the law, abortion amounts to homicide, if not actual murder. The former says that any act committed to hide a sin, is itself sinful. Now, if conception out of wedlock is sinful, then, a child so conceived should be regarded as an embodiment of sin—a living monument to sinfulness. It should not be allowed to soil the moral atmosphere of society. Therefore, to kill such a child at birth or to prevent it from being born, should be encouraged by the custodians of social morality.

The legal attitude towards abortion is utterly incomprehensible. Until the child is born, it has no existence, socially. So, the question of murder or homicide does not arise at all. An act concerning something which is not a part of society, cannot be interpreted as an offence against society; therefore, it does not come under the purview of laws which are meant to punish offenders against society. Moreover, until it is delivered, the child is a part of the mother's body—her property in the physical as well as in the productive sense. Ownership carries with it the right of disposition. Law does not punish a woman for burning to ashes a pearl necklace legitimately belonging to her. But it puts her in prison if she chooses to destroy an embryo in her womb which is her private possession more intimately than the necklace. Is it not senseless? It is as ridiculous as to punish one for inflicting a wound on his body. Before the child acquires the double position—child of its parents and an incipient social unit—it entirely belongs to the mother. Therefore, by punishing abortion, criminal law contradicts its own fundamental principle. It penalises the exercise of the right of ownership precisely in such a case wherein this is acquired naturally, and therefore inalienable. Before the child is born, it does not exist socially; so, society has no call to take it under protection. How do you know it wants to be protected? Besides, when it really needs protection, society seldom comes to its aid. Who punishes society for letting countless children die of malnutrition, lack of medical aid, bad hygienic conditions and other preventable causes? Society officiously takes embryonic lives under its protection, so that it can kill them wholesale when they have become human beings. Governments commit murder *en masse*, in various ways. War is the most glaring example. Is it not sheer impudence for such Governments to send individuals to the gallows for a "crime" which pales into utter insignificance when compared with the crimes committed by themselves? And to punish, abortion? The unreasonableness is glaring.

If abortion is a crime, the social system that compels its commission is responsible for it. No mother would ever destroy her child, unless she is absolutely compelled to do so. The compulsion results from the convention which makes motherhood dependent on wedlock. Motherhood is a gift of nature. Marriage is a man-made institution. The consummation of a natural impulse is branded as immoral, if it does not take place within the limits of a man-made institution which does not make room for all. Under no condition, motherhood can be immoral. On what moral ground can a widow be deprived of the natural right to motherhood? Nor is there any reason to brand the unmarried mother with the stigma of sinfulness. Naturally and morally, no child is illegitimate. Recognise this simple truth, and the tragic practice of abortion will disappear. A mother reluctantly destroys her child, only when she has to be ashamed of her motherhood instead of being proud of it. The anxiety to avoid the shame of illegitimate motherhood, leads to the commission of the crime of abortion which, committed necessarily clandestinely, with crude, often dangerous, means, under insanitary conditions, involves the life of the mother as well. Sometimes the mother is killed in order that the new life germinated in her "sin" may be destroyed. Obviously, that was what happened in the case here recorded.

The harassed old woman certainly did not want to kill her daughter, however much she might have disapproved of her misdemeanour. But steps taken for sparing the family shame and social persecution, led to the death of the unfortunate girl. The mortification of the poor old lady could be easily imagined. As if she was not sufficiently punished by the tragic consequence of her own act, done in deference to a callous society and unthinking compliance with a perverse code of morality! In addition to that, she had to pay the penalty of law. Helpless victim of cruelty and injustice, she blamed her fate, and sought consolation in her faith in the doctrine of *karma*.

VI

There was the case of an entire family punished on the charge of murdering one man. Altogether six, including two women. The father with one of his three sons was hanged. The remaining two brothers, the mother and a daughter-in-law — all given transportation for life.

The murdered man was the village moneylender. The family punished for murdering him lived on a few bighas of land, and was heavily indebted as all poor Indian peasant families are.

The meagre means of subsistence of a number of hard-working people was at the legal disposal of the money-lender who held a mortgage on it.

The charge was that one day the money-lender was invited to the house of the debtors who proposed to do business over a friendly meal. It was alleged that the plan was to poison him; but somehow he survived the murderous plan which, granted that the allegation was true, by its very nature, must have the criminal co-operation of the women-folk of the household. The would-be murderers, however, were determined to dispatch the benevolent bania out of the arduous life of collecting compound interest from a lot destitute of the means for meeting their legal obligations in full, and promptly on demand. It was argued from the side of the prosecution that the shrewd bania had smelled rats, and left the friendly feast rather precipitately; the father of the family with one son accompanied him, and hacked the good samaritan in a wood on the way The argument was that the behaviour of the guest had frightened the criminal conspirators who realised that, to head off sure exposure of their murderous plan, they must finish the job somehow; and they did with *dao* what could not be done neatly with poison.

The defence was that the distressed debtors had hoped to mollify the money-lender with hospitality, and then persuade him not to drive them off the land as he had threatened to do if they failed to pay off in full. But with all the Indian's spiritual view of life, the bania had behaved like the materialist American, and insisted upon doing business first. He was alleged to have made scornful remarks about the proferred hospitality—that instead of wasting money in feasts, honest and god-fearing people should meet their legal obligations; that good people would rather starve than deny their legitimate dues. The hosts had resented the attitude of the guest, and there had resulted an altercation. Upon that, the bania had left in a rage, saying that the invitation was a plan to poison him, and that he was going to report it to the police.

Terrified by the ominous threat, the father followed by one son, had gone with the indignant man with the object of pacifying him, so that he would not carry out his threat. Instead of being pacified, the enraged bania had grown more insulting, and used abusive language with reference to the women. Since one of these was the young man's wife, he had lost his temper and assaulted the vilifier who was severely wounded in consequence. Afraid that, if they left him in that condition, he would surely do the greatest possible harm, the father had aided the hasty son with the hope of finding safety in the belief that a dead man tells no tales. So, as far as those two were concerned, the murder was confessed, and law sent them to the gallows with a clear conscience. The rest was sentenced, and given the maximum punishment, on circumstantial evidence.

Granted that the prosecution case was properly proved (it must have been, technically, otherwise a competent Court would not pronounce the judgement), and that the defence story was unreliable, what about the facts that, though not mentioned in defence, constituted the setting in which the crime was committed? Itself an instrument for maintaining the social *status quo*, criminal justice, nevertheless, does not admit in evidence facts that are contributory to, often compel the commission of, crimes it punishes. Social causes of crime are dismissed as irrelevant to the administration of criminal justice which is an act of social coercion.

However, extenuating circumstances are taken into consideration even when the technical responsibility for the crime is proved. Was there no extenuating circumstance in this case? At least as far as the four not directly involved in the murder were concerned, particularly the women?

But the point is not this or that particular case, or punishment of individuals. It is that here we have a type of "crime", the responsibility for which clearly belongs to the socio-political system that punishes it. The law is not a purely political instrument; that is to say, governmental affair, as it is generally believed to be. It has a social sanction. The Indian society cannot shirk the responsibility for the legal punishment of this type of crime with the convenient plea that it had no hand in the making of the law. Didn't it? That is the question. Whence comes the shameless opposition to legislative measures meant not to harm the money-lender, but only to give some very inadequate protection to his helpless victims? Even to-day, many Indian public men vehemently disapprove of projects for a radical cure of the cause of mass indebtedness which places people in such a hopeless position that they are compelled to commit crimes for the sake of remaining in human existence, supposed to be a gift of God. They want to help the peasantry, but would not do injustice to other classes. The peasants are to be helped out of their ruinous indebtedness, but the "legitimate" right of their usurious creditors should also be respected. The nationalists are the spokesmen of the Indian society. They are opposed to the present Government; but they approve of the principle of law in Courts. The principle is coercion for the maintenance of the social *status quo*.

Opponents of any radical measure maintain that the rural masses themselves are largely responsible for their indebtedness. The most oft repeated argument is that they get into deep waters owing to the habit of spending beyond their means in ceremonials such as marriage etc. In the first place, though the charge is not altogether groundless, it ignores the fundamental fact of the economic condition of the toiling masses either deliberately or through naive disregard for realities. The fact is that such a large portion of the peasants' net income is taken away from them, in one form or another, as does not leave them, in the great majority of cases, nearly sufficient for bare existence. Practically the entire bulk of the peasantry is always short of working capital, and is obliged to borrow it. This fundamental fact of the situation not only exposes the bankruptcy of the peasant masses, but vitiates the whole system of national economy. It is so very evident that to-day none dare deny it. There are, of course, all sorts of explanations; but the fact remains, not to be explained away. The disease eats into the vitals of the entire social structure, and cannot be cured by quackery, whether preached by prophets who would make omelette without breaking even one single egg, or practised by a niggardly Government whose hands are tied by its own nature. The latter's efforts are bound to be as ludicrously ineffective as the effort of pulling oneself up by his own boot-strings. Meanwhile, crimes of all kind must spread as ugly symptoms of a chronic disease. The legal punishment of these crimes is morally as justifiable as penalising the victims of a contagious disease.

In the second place, granted the contention that the indebtedness of the masses is to some extent due to the expenditure on ceremonials, the seekers after the root-cause of the disease must ask: Whose fault is that? If those, often without the means of bare subsistence, are found as a rule to incur occasional expenditures which aggravate their plight, already hopeless, they must be acting under compulsion. Expenditure for ceremonial purposes are socially compelled. They are seldom made voluntarily. Therefore, the responsibility on that account cannot be laid at the door of the people who practise them. It belongs to religious legislation, social custom, and cultural tradition.

Religion teaches that it is the duty of children to perform certain ceremonies on the death of their parents, the default in which would not only cause inconvenience to the dead, but prejudice the post-mortem position of the defaulters themselves. These ceremonies, prescribed as religious practices, cannot however be performed in a simple religious manner—through prayer or worship, for example. They are rituals, and as such require more than a spiritual attitude. They are essentially means for paying material tribute to the administrators of the religious law.

The superstition that the deliverance of the soul of the dead is conditional upon the due performance of certain prescribed rituals having been traditionally ingrained into the popular mind, the expenses necessary for the purpose are matters of obligation, though met voluntarily by the superstitious. To hold them responsible for this ruinous habit, is no more reasonable than to rebuke a drug-addict after having forced him to cultivate the pernicious practice. Like Government, religion also takes tribute; and the greater the ignorance of the faithful, the more the tribute exacted for the protection of their souls. For the Indian masses, the laws of Manu or of the Sheriff or any other religious law, not only religiously administered, but often fraudulently interpreted by the custodians of the spiritual law and order, is more binding than the temporal laws of the political regime. These they obey out of fear; whereas religious laws, though no less, indeed more, ruinous economically and coercive socially, are obeyed by habit which, born of ignorance, has been skilfully cultivated through ages by the beneficiaries thereof, and is deeply entrenched in superstition fostered fanatically by the ideologists of the forces of reaction which thrive parasitically on the material misery and moral degradation of the masses.

Economic bankruptcy of the masses, progressive pauperisation of the poorer strata, is a fertile breeding ground for crimes, the responsibility for which, therefore, belongs to the socio-political system which cannot exist without creating those deplorable conditions. Crimes are symptoms of social malady. The only effective and moral manner of combating them is to get into grips with the disease, the nature of which is indicated by them. As long as there will be people driven to despair by indebtedness, inevitable because incurred under pressure of forces beyond their control, crime of the nature described above cannot be checked by any number of exemplary punishments as inflicted as in this case.

VII

Finally, there is the case of a man hanged for massacring his whole family. He was regarded as a maniac. Yet, instead of being placed in a mental hospital, the poor fellow was sent to the gallows, in strict accordance with the dictates of merciless justice. God is just and merciful; but justice, which is supposed to be an attribute of God, is merciless. Queer logic!

The crime committed by the man could be described as atrocious; nor did he deny the charge. He could not, having done his murderous deed quite openly. That is why he was described as a maniac. But his explanation of the extraordinary behaviour was highly interesting.

For some time, his wife had been regarded as a witch by the inhabitants of the village. In the beginning, he had not paid much attention to the story whispered about. But gradually, it became a matter of general belief, and he was approached by the village elders with the demand that he should drive away the mother of his children, whose presence was believed to be harmful to the village. The charge was that she could bring about the death of anybody by simply wishing it, and that a number of untimely deaths had recently occurred on that account. It is surprising that the man, himself an ignorant rustic, did not readily fall victim to the popular superstition, and do as demanded of him. It would be quite natural for him to behave so. But, to the utter chagrin of the village *vox dei*, he behaved otherwise. He refused to act on the noble model of Ramachandra; he would not sacrifice his wife, the mother of his children, to placate vulgar superstition.

For his kind, that was an incomprehensible attitude. There was no ground to assume that the villagers were moved by malice, although it could not be altogether excluded that the charge against the woman had been the result of some private animosity. However, once the rumour was afloat, the villagers generally came to believe sincerely that the woman was a witch. Sincerity, that is to say, honesty of faith, is the most fundamental characteristic of superstition. That is why superstition dies so very hard. They being honest about their belief in the evilness of the woman, it was incomprehensible for them how someone could fail to share it. The relation between the non-conformist and the persecuted woman was a matter of no consequence. What is a wife? One does not place the loyalty to a mere woman above popular suspicion, based on traditional faith. Even Ramachandra did not do so. He exiled Sita to placate public opinion. The unprecedented behaviour of the man aroused suspicion against himself. He was believed to be an accomplice of his wife. He could not possibly stand by the evil woman, unless he was himself evil. Before long, there were stories about co-operation between the man and his wife in various nefarious acts. They were shunned by all the villagers who were afraid of them. The entire family came to be regarded as a brood of evil.

It was not a very pleasant position to be in. What could the poor fellow do? The whole village was scared out of wits; and the neighbourhood shared the superstition. Fear deadens reason, the very little of it possessed by the superstitious. He could not persuade the people to see that they were scared of a spectre invoked out of their credulity. Nor was he free from superstition himself. He would have been with the crowd, had the object of persecution been someone else. He did not have anything like a rational conviction to fall back upon as the unshakable line of defence against the onslaught of popular prejudice. He was thoroughly bewildered by the turn of events. It was too late to throw the accursed woman overboard which he would have done to extricate himself and his family from the intolerable situation. The only possible way out would be to leave the village, and go away beyond the reach of the gratuitous notoriety. But even that did not hold out any convincing promise. For, he was no longer sure that his wife was not what she was suspected to be. Superstition would not permit him to see that, as he was believed to be a wizard, though he was not, so might also be the case with his wife. If his wife was really a witch, and he had come to share the general fear, what was the use of going away?

To take her along would be to remain wedded to the misfortune. So long as the loadstone hung around his accursed neck, he was sure to sink, no matter where he went. On the other hand, the idea of leaving her behind was not appealing. He was not sure that she could be so easily got rid of. Fate had placed him in a position which had no way out. There was a more potent consideration which precluded the possibility of escape. Where could he go? How should he maintain his family in a new place? All his earthly possessions consisted of a bit of land. That he must leave behind. So, should he leave the native village, with or without the impediment of the witch of a wife, the only destination would be eventual death from starvation. That was not a very alluring escape. It is in human nature to cling to the physical being, however unbearable, until it becomes psychologically impossible. Suicide is always committed in a state of temporary insanity. Despair creates the pathological condition.

The man was actually driven mad by the helpness of his position. The crime for which he was punished, gruesome as it was, was an act of madness. The man was killed with the ceremonious sanction of law, not for any fault of his own, but because popular ignorance, fostered by our spiritual culture, drove him mad. In any case, how is society benefitted by law adding murders to those committed without the sanction of law? How is it justifiable for law to commit the crime it presumes to punish? Justice is still far from being civilised. She remains as avage Goddess who demands human sacrifice. And gods and goddesses are made after the image of their worshippers. If the world was really civilised, it would not worship a savage goddess with the offering of blood. You may place the offering in an electric chair; yet, it is human sacrifice.

VIII

At the last moment, a very singular case came to my notice. Being of a different type, I must record it.

In this case also, the crime is murder, committed by a young man who was sentenced for the offence to transportation for life. The victim was a religious mendicant. The judge had no doubt about the guilt of the accused; but he generously thought that it was a case in which justice might be tempered with mercy, because the crime was committed under grave provocation. The nature of the provocation makes the case interesting.

The culprit belongs to the class of wage-earners. Having failed to find any employment in his home town, he went to distant parts in quest of it. But in these days of wide-spread unemployment, one place is as bad as another. Consequently, the young man (he was rather a boy, being still in his 'teens) wandered from place to place, becoming, like others without number, sort of a vagrant. In course of his peregrinations in search of employment, not to be found anywhere, he happened to pick up acquaintance with the mendicant.

It is a peculiarity of Indian society that, while those willing to perform productive labour for their daily bread starve in hundreds of thousands, religious mendicancy is a profession which supports myriads of parasites. The vagrant boy told his story to the holy man, and was asked to be his *chela*, in which capacity his daily bread would be provided for. Being hopeless to find any employment, the boy accepted the generous offer, and became an apprentice in the only profession which still pays in India. But there was a condition in the contract which, though not stipulated by the holy man in the beginning, was imposed on the other party subsequently. The boy seems to have submitted for a time reluctantly to the condition which he found to be odious. Finally, he rebelled, and killed the holy man for exacting a repellant price for measely morsels of begged bread.

The boy pleaded not guilty to the charge. The body of the murdered man was found in a lonely place, and several villagers testified that they had caught the boy running away in blood-stained clothes. The story of the accused was that he had gone away for a while leaving the holy man asleep in the lonely place. On his return, he found him lying frightfully wounded and rushed to his aid. Thereupon, some men appeared on the scene and caught hold of him.

The interesting piece of evidence, which indicated the motive of the crime, was that apart from other wounds inflicted with a stout branch the sex organ of the murdered man was smashed up. It was clear that this was not caused by any of the blows that killed the man; nor would any chance murderer select such a part for dealing blows with the purpose of killing the victim. Evidently, the tell-tale blow was delivered after the man had been killed or overwhelmed, for ventilating the repugnance to certain behaviour of the holy man which had supplied the motive of the murderer's act. The judge himself seized upon that interesting piece of evidence which, in his legalistic opinion, established the guilt of the accused, in spite of the circumstantial nature of the entire evidence against him, and, at the same time, showed that he had acted under grave provocation, thus entitling him to leniency of punishment.

There is no doubt that the boy killed the holy man; and it is also clear why the murder was committed. The question is: *morally*, who was the criminal? If one must pay the penalty for a crime committed, did not the murdered man only get his dessert? The punishment was excessive? Well, that, deliberately, on sober consideration, inflicted on the boy, was no less so. The boy was admittedly driven to the crime. It might be argued that the boy could have left the holy man if he did not like complying with his demand. But that would simply be condoning the hateful behaviour of the holy man. As if he was morally justified in making the demand, because he gave the boy his daily bread. Besides, utter destitution had forced the boy to the easy profession of mendicancy. Naturally, he would be reluctant to forfeit the privilege of parasitism derived from the association with the holy man. It was morally perverse on the part of this latter to exact such a price for the privilege. It would be a different matter, had the boy voluntarily accepted the condition from the beginning. Evidently, he did not. Afraid of the dire prospect of returning to the hunt for employment, never to be found, the boy submitted to what he loathed. While reluctantly complying with the repugnant condition in payment for the privilege of eating the bread begged by the mendicant, the boy came to hate the man. The murder was the fierce expression of his hatred which the latter fully deserved.

This case reveals the practice of moral perversity that is usually hidden behind the much vaunted virtue of celibacy. The profession of sexual continence leads to abnormal practices. Nature cannot be cheated. The vain notion that the religious can rise above the law of nature only breeds self-deception and hypocrisy. The few who really practise as well as profess continence, become psychologically deranged, given to morbid fantasies, and hallucinations (trance, etc.) glorified as tokens of spiritual elevation. These expressions of spirituality are distorted emotional states brought about by suppressed desires. Nature takes vengeance.

However, by and large, celibacy is a sham which gives rise to all sorts of licentious and perverse practices, in their turn often leading to the commission of crimes. Would not the holy man be less despicable morally if he had visited the house of a prostitute in order to satisfy his natural desire, instead of doing so in the abominable manner he adopted? Even homosexuality, as such, need not be regarded as moral degradation. Preferred to normal intercourse, it is rather an emotional morbidity to be treated psychiatrically. But with those like our holy man, it is not a matter of preference, but practised because it is cheaper and can be done clandestinely with greater facility. Thus, it must be regarded as sheer moral perversity, and those given to it condemned as corrupt criminals, masquerading in a religious guise. Rational beings could hardly adhere to, and proudly defend, a religion that encourages such corrupt and corrupting practices under the veil of venerable professions.

Not only in India, but everywhere, crimes are symptoms of social disease. They have been that throughout the ages, all over the world. And instead of treating them as such, attempts have been made to suppress them more or less cruelly with the sanction of religion, ethics, social codes and law. The attempts have failed as they must. So long as society remains based upon ignorance, superstition, brutalisation, suppression, poverty and degradation of the masses, so that privileges of caste, sex, class (as the case may be) can be preserved, crime cannot be successfully combated, however severe may be the punitive measures, or well organised the administration of justice. The more decadent a society, the uglier and more wide-spread are the crimes. Forms of crimes are determined by the social conditions under which they are committed. While the majority of criminal acts are economically motivated, in India as well as in any other country, socially organised as described above, the peculiar cultural atmosphere of this country breeds specific types of crime. There is still another peculiarity, which springs from the same source.

In every country, legal punishment of crime is sanctioned by the established religion and orthodox codes of morality. But as a rule, the criminal, while bowing down before legal justice, enforced by the power of the State, is more or less rebellious as regards the religion and morality which endorse his punishment. In India, the criminal's psychology co-operates with his punishment. Even if he feels that law has done him injustice as he often does, victimisation of the legally innocent being not infrequent, he himself supplies the moral justification of the injustice done to him. It is all written in his fate,—result of his *karma*. Instead of revolting against the injustice, he finds in it divine dispensation.

Fatalism, which sums up the much advertised spiritual nature of the Indian people, teaches him to hold himself responsible not only for the social conditions that have driven him to crime, but even for the legal injustice palpably done to him.

I know many a man serving long terms of imprisonment for crimes they did not commit. Yet, they are totally devoid of any moral indignation or even resentment. If the facts of social iniquity and legal injustice are pointed out to them, with a religious fervour, they touch their forehead. It is all unalterably predetermined; no use complaining against it. Indeed, it is sinful to do so. They are reconciled to the hard lot, accepting it as divine dispensation. Such supine resignation to injustice is not a token of spirituality. It is moral degeneration; and this shameful psychological state of the Indian masses is our cultural heritage. Is it anything to be proud about?

Here we find religion standing naked as an instrument not only of traditional and time-honoured social oppression, but even of crass legal injustice. Religiosity has broken the backbone of the Indian people. It has crushed the spirit of revolt, which is the lever of human progress. Resignation, the characteristic feature of the Indian people, leads to stagnation which is living death. Thousands of victims of social iniquity, and high-handedness of the minions of law and order, finding solace in the religious superstition of fate or *karma*—that is the depressing sight to be seen in Indian prisons. And the jail population being a cross-section of society outside, the sight presented by it transcends the prison walls, and provides food for serious thought to those who have the courage to distinguish facts from fictions, to look into the frowning face of rude reality.

A study of the psychology of the Indian criminals throws a flood of light on the doctrine of *karma*; it enables one to appreciate how spiritually corrupting and morally degrading the religious view of life is. Religion is based on ignorance. The future of India is seriously prejudiced by her cultural traditions which keep the masses in ignorance, so that superstition can remain alive as the foundation of the religious view of life. In so far as the masses are concerned, our religious or spiritual culture stands for ignorance, superstition and fatalism. There is no future for a nation, the vast bulk of which remain in such a deplorable state of spiritual and moral backwardness. How can it be free when its cultural tradition teaches people to make a virtue of submission? And the most curious thing is that even the prophets of national freedom are fanatical, though not always sincere, protagonists of that very tradition which precludes the will to freedom. Thanks to this curiosity, the movement for national freedom has been caught up in a vicious circle of futility and impotence. Revolt and submission are mutually exclusive. The first condition for national freedom is to break away from the paralysing grip of the hands of the dead past. Loyalty to the dead often amounts to betrayal of the living. The captive of the past can never conquer the future.